MATERIALS SCIENCE AND TECHNOLOGIES

RADIO FREQUENCY IDENTIFICATION

Materials Science and Technologies

Additional books in this series can be found on Nova's website under the Series tab.

Additional E-books in this series can be found on Nova's website under the E-books tab.

Engineering Tools, Techniques and Tables

Additional books in this series can be found on Nova's website under the Series tab.

Additional E-books in this series can be found on Nova's website under the E-books tab.

MATERIALS SCIENCE AND TECHNOLOGIES

RADIO FREQUENCY IDENTIFICATION

ALISON R. MCADAMS
EDITOR

Nova Science Publishers, Inc.
New York

For permission to use material from this book please contact us:
Telephone 631-231-7269; Fax 631-231-8175
Web Site: http://www.novapublishers.com

NOTICE TO THE READER

Additional color graphics may be available in the e-book version of this book.

LIBRARY OF CONGRESS CATALOGING-IN-PUBLICATION DATA

Radio frequency identification / editors, Alison R. McAdams.
p. cm.
Includes bibliographical references and index.
ISBN 978-1-61122-416-0 (hardcover)
1. Radio frequency identification systems. 2. Inventory control. I. McAdams, Alison R.
TK6570.I34R33 2010
658.7'87--dc22
2010041306

Published by Nova Science Publishers, Inc. † New York

CONTENTS

PREFACE

Radio-frequency identification (RFID) is the use of an object applied to or incorporated into a product, animal, or person for the purpose of identification and tracking using radio waves. RFID has many applications; for example, it is used in enterprise supply chain management to improve the efficiency of inventory tracking and management. This book presents current research from across the globe in the study of radio frequency identificatio, including the role of RFID in agriculture; plants with implanted RFID microchips to yield safer and more wholesome products; radio frequency identification in the support and transmission of medical information in the field of disaster medicine; and RFID adoption in the developed and developing world.

Chapter 1- Passive radio frequency identification (RFID) technology has several limitations due to physical constraints in signal transmission and data processing in the supply chain. Therefore in order to access more information about the product, it is necessary to store that information externally. However, standard compliance RFID infrastructures are expensive. Many enterprises, large and small, are struggling to invest in this emerging technology and are left without RFID connectivity. Hence, some parts of the information link of the supply chain are broken. This chapter draws upon the experience of two large scale pilot projects using electronic product code (EPC) technology to illustrate the issues of connectivity in RFID implementation. The chapter then describes a new expanded supply chain infrastructure system that adopts a generalized RFID data containment capability that incorporates other mobile and more cost effective technologies.

Chapter 2- RFID is entering in a new phase. RFID technologies are said to improve the performance of many agricultural processes. Recent advances offer vast opportunities for research, development and innovation in

Agriculture. This is the consequence of lowering costs of ownership, engineering increasingly smaller sensing devices and the achievements in radio frequency technology and digital circuits. The aim of this chapter is to give readers a comprehensive view of the possibilities of RFID application in agriculture.

RFID was originally developed for short-range product identification, typically covering the 2 mm - 2 m read range and has been successfully applied to food logistics and supply chain management processes. With an increasing demand for security and safety, complete documentations for food products, from field to customer, have become increasingly demanding. RFID has been accepted as a new technology for a well-structured traceability system on data collecting, and human, animal and product tracking. Also, electronic identification of cattle using RFID is a common practice in many farms. These devices are used to monitor animal behavior in mid-size outdoor pens, providing digital data that can be easily computerized. However, recent developments in RFID hardware outfitted with sensors extend its range of application. There are commercial active and semi-passive tags that can collect temperature information. Other semi-passive tags outfitted with sensor are under development, like humidity, shock/vibration, light, pH and concentration of gases, such as acetaldehyde or ethylene. Moreover, the last generation of Class 4 RFID tags can be configured in a mesh network. In this type of network, the tags can communicate each other to get to a reader circumventing environmental obstacles and extend the size of the system.

These specialized RFID monitoring devices promise to revolutionize a wide range of agricultural operations. The development of applications in precision agriculture, like monitoring crops, makes possible to increase efficiencies, productivity and profitability while minimizing unintended impacts on wildlife and the environment, in many agricultural production systems. Instead of take decisions based in some hypothetical average condition, which may not exist anywhere in the reality, a precision farming approach recognizes differences and adjusts management actions accordingly.

In cold chain of perishable food products RFID monitoring provides new features that have the potential to be an economically viable, alerting if the products are not stored at the right temperature and predict the remaining shelf life. In this field, several applications for monitoring cold chain logistics by means of RFID have been reported.

Chapter 3- In recent years consumers have demanded safer and more wholesome products, and stricter regulations have supported these expectations. Thus, the need to know more about the origins and qualitative characteristics of food products commercialized worldwide has increased. Traceability is essentially the "history" of a product, from its origin to the shelf, and in agriculture traceability reaches back to the genetic status of products, whether they are of animal or vegetal origin. In order to verify the status of a product, different tools are used during different phases of the production line.

In agriculture, radio-frequency identification (RFID) technology has been introduced efficiently in animal identification systems for traceability purposes, and legal regulations have been put in place for this sector. Differently from animals, where the microchip is often inserted within the organism, RFID applications for plants mainly regard food traceability, logistics or harvesting, in which the microchip (tag) is not inserted inside the plant or product.

In the last 5 years, experimental trials focusing on inserting tags within plants have been carried out on *Citrus* spp., *Cypress* spp., *Platanus* spp., *Vitis* spp. and other genus. Keeping in mind plant histology and organ size, different techniques and tag allocations have been proposed, yet the standardization of RFID tagging in plants does not seem possible. In fact, specific solutions have been suggested for tagging plants with respect to growth stage, anatomy or aesthetic considerations. Moreover, even the aim of tagging can orientate the technology used and, consequentially, the methods of tag insertion. Nowadays, identification of mother plants used for propagative purposes or plant pathology monitoring represent the most relevant cases of study or practical applications, but there are also some interesting outlooks with regard to RFID integration with precision farming and for developing information management systems.

Implanting an RFID microchip inside a plant represents the first step for developing integrated systems that involve positioning techniques, mobile or Wi-Fi devices and a Web 2.0 approach. These complex systems can strongly support farm management, differentiate the final product in markets, and help controllers in their verification and increase consumer confidence in product origin.

Chapter 4- Support and transmission of medical information are still challenging the field of Disaster Medicine.

The department. of SAMU (Service emergency medical assistance) of Lyon (France) and its unit of Disaster Medicine are continuously looking for new technologies to enhance medical doctors actions and ensure traceability.

The authors therefore tried to apply the RFID method in this context, and to computerize the medical chain of relief at the scene of a disaster.

The authors will at first define the so-called situation of a disaster, and the role of French SAMU. Second, they will describe the management of these situations as planned by the Government in the red channel. The authors will then give details of the means and methods for computerization of these emergency resources, with particular attention to the use of RFID.

Chapter 5- Radio Frequency Identification (RFID) has gained international recognition based on many potential benefits it possesses. However, RFID adoption is facing a lot of challenges that many new and emerging technologies face. Some of the challenges peculiar to RFID adoption are cost, standardization, legality, privacy, time, fear and so on. This paper explores the challenges and benefits of adopting RFID in almost everyday life applications. This paper also discusses and gives recommendation on how to bridge the gap between the developed and the developing World in regards to RFID adoption.

Chapter 6- In this paper the authors have overviewed the latest technology RFID (An automatic identification procedure) by comparing it with the ongoing technologies like barcode, smart cards on various parameters. There is a brief chronological history about the development of this technology starting from 1800 to till date developments. The authors have also discussed about the essential features of the RFID system, which are tags, readers, application software. They then have illustrated in detail various types of tags:-Active, Passive, then about readers and how the tag and reader are coupled. These mainly work at 13.56MHz frequency, under international standards. The most important advantage of RFID is its data security which is also being dealt with along with its application, advantages and disadvantages. This technology is widely being used in different fields like health, animal identification, passports, making it a ubiquitous identity card for every fields, libraries etc.

Chapter 7- Radio Frequency Identification (RFID) technology has received overwhelming attention in recent years. RFID tags could be regarded as ubiquitous or pervasive computing devices, which can send or receive radio frequency signal and can store a number of data. It is in fact not a new technology because its first military application can be found in the Second World War. Recently, RFID-based systems are linked to multi-agents systems such that RFID-based devices are considered as agents, in particular mobile agents. However, reported literature is related to, relatively, high-level conceptual applications of RFID technology in agent-based systems. The answers to the following questions are still unclear: whether RFID-based systems could be represented as multi-agent systems or not? If this is not true, what are the missing components in RFID-based systems in contrast to multi-agent systems? In other words, how RFID-based systems could be enhanced to become multi-agent systems? The objective of this chapter is to address these questions.

In: Radio Frequency Identification ISBN 978-1-61122-416-0
Editor: Alison R. McAdams, pp. 1-30

Chapter 1

RFID INFRASTRUCTURE FOR GLOBAL SUPPLY CHAINS INVOLVING SMALL AND MEDIUM ENTERPRISES

John P. T. Mo
RMIT University, Australia

ABSTRACT

Passive radio frequency identification (RFID) technology has several limitations due to physical constraints in signal transmission and data processing in the supply chain. Therefore in order to access more information about the product, it is necessary to store that information externally. However, standard compliance RFID infrastructures are expensive. Many enterprises, large and small, are struggling to invest in this emerging technology and are left without RFID connectivity. Hence, some parts of the information link of the supply chain are broken. This chapter draws upon the experience of two large scale pilot projects using electronic product code (EPC) technology to illustrate the issues of connectivity in RFID implementation. The chapter then describes a new expanded supply chain infrastructure system that adopts a generalized RFID data containment capability that incorporates other mobile and more cost effective technologies.

Introduction

A global supply chain has many types of participants including manufacturers, logistics companies, distributors, retailers and government agencies. An important success factor for a supply chain is the ability for the participants to track products moving through the supply chain [1]. Such systems already exist for a few decades in the form of "Scan Pack", which transfers trading information via EDI (Electronic Data Interchange). EDI works by sending order information electronically to the supplier and/or importer, or other trading partners [2]. Additional product identification data including Serial Shipping Container Code (SSCC) labels and Advance Shipping Notice (ASN) are created. These data are scanned many times in the shipment transactions by the supplier, importer, distributors, retailers, and so on. The process requires a lot of manual handling and is prone to errors. The cost in managing the Scan Pack process is therefore very high. Furthermore, most Scan Pack systems are standalone applications. They are difficult to be interfaced to enterprise management systems [3]. Furthermore, many supply chains are operating globally. Goods are manufactured and scanned in countries where IT infrastructures may not be compatible with the receiving countries. New technologies are required to overcome these data integrity issues.

Radio Frequency Identification (RFID) emerges as a re-vitalized technology from invention dated back to World War II. RFID uses radio frequency to capture data from RFID tags (also known as RFID transponders) attached to products as they are moved through the supply chain. The radio frequency signal is read and interpreted by a RFID reader (also known as RFID interrogator). The key advantage RFID has over other forms of identification technologies is its ability to be read without line of sight. If the protocol for tag reading is managed correctly, multiple tags can be read simultaneously. These abilities give huge potential to the supply chain to increase efficiency and productivity [4]. Two types of RFID tags are used. A passive tag energizes its circuit by energy captured from the RFID reader from its antenna. The cost of passive tags can be as low as a few cents each. On the contrary, an active tag is powered by its on-board battery and hence can transmit or read from longer ranges. Passive tags are commonly used in supply chain but are not reliable due to physical constraints [5]. In order to access more information about the product, it is necessary to store that information externally. For most consumable product supply chains, to find information such as production date, batch number, package size, the identity of an RFID tag is used as a key to access product data from a server [6]. However, storing these data in a single location managed by one of the supply chain participants is not very useful due to the fact that the tags will need to be read by several parties [7]. Therefore it is common practice to store the data in a secured distributed environment, where management of data is independent of any supply chain partners, but allowing ease of access to the authorized parties.

In the past few years, many researches investigated the benefits of RFID application to global supply chains. One of the important developments is the Electronic Product Code (EPC) standard defined by EPCglobal (http://www.epcglobalinc.org/). In an analysis of the economic impact of the EPC system to the Fast Moving Consumer Goods (FMCG) supply chain, it was found that not everyone could benefit from its implementation [8]. In warehousing, RFID has been used for indoor item tracing, supporting storage and retrieval activities [9]. In US, suppliers responding to Wal-mart's RFID mandate experienced net gains in abnormal stock returns [10]. In Australia, two national demonstrator projects were conducted to investigate the effectiveness of applying EPC to assist FMCG supply chains to improve their efficiency. The National Electronic Product Code Network Demonstrator Project (NDP) tracked pallets and cartons through the supply chain [11]. The NDP Extension Project explored the issues related to the application of EPC data to facilitate business transactions [12]. In Europe, the project "Building Radio frequency IDentification solutions for the Global Environment" (BRIDGE) was developed to resolve the barriers to the implementation of the EPCglobal Network [13].

In order to be competitive, companies participating in the global supply chain have been implementing strategies that make them highly integrated within their own individual's boundaries [14]. However, the side effect is that it prevents inter-enterprise systems integration [15]. Practices and regulations are different in different countries. More importantly, many supply chain participants, particularly small to medium enterprises (SMEs) do not have resources to explore and implement RFID technology. It is necessary to develop a system that works across companies, organisations and country boundaries. A systematic modelling method is required to understand and plan for the functions, activities, work flow and processes in the scenario [16]. The use of enterprise modelling methodology will ensure a unified view among supply chain partners.

An enterprise model represents business systems in a common perspective that is shared among different groups and enterprises. The IFIP–IFAC Task

Force [17] specified a generic enterprise reference architecture that defined all virtual enterprise operating elements including role functions, project execution guides, document repository, documentation, work flow control and entry portal. Analysis of these elements helps the enterprise engineer to identify structural issues in the enterprise and resolve them by modifying the enterprise design. This chapter reviews the outcomes of two national demonstrator projects in Australia and examines the enterprise models that drive the processes in the demonstrators. The impact of these models to the infrastructure systems of the supply chain are discussed in relation to future identification technology requirements.

The National EPC Network Demonstrator Project

The National EPC Network Demonstrator Project (NDP) aimed to identify the business benefits of sharing information securely using the Electronic Product Code (EPC) Network for tracking product movements among multiple enterprises. This project demonstrated the principle of the EPC network enabling inter-enterprise transactions and supply chain management. When a given tag was detected, instead of having each company storing this information and communicating to all related partners, the RFID infrastructure defined a unique global EPC that could be queried for links to access detail information from local servers.

The NDP was a large scale project with 13 consortium members between two cities in Australia. To ensure adequate resources for the project, the project focused on tracking 9 product items incorporated in 15 RFID handling processes [18]. Several innovative process designs were developed to support data integrity of the system.

Pick Face Process

One of the fundamental business processes within the project was the building of a pallet of cartons. The consortium considered that the possibility of a RFID reader detecting all (100%) tags within its RF field at a read point was low. It was necessary to ensure that the manufacturer knew exactly what had been packed in a shipment or on a pallet at some stage in the supply chain.

The "pick face" process was designed to achieve absolute confidence on the identity of the item. The process required the registration of cartons to the EPC system (Figure 1). After the registration, the cartons were "picked" onto the pallet in a controlled fashion. When the pallet was full, the process was concluded by "closing" the pallet, i.e. no more items should be placed on the pallet.

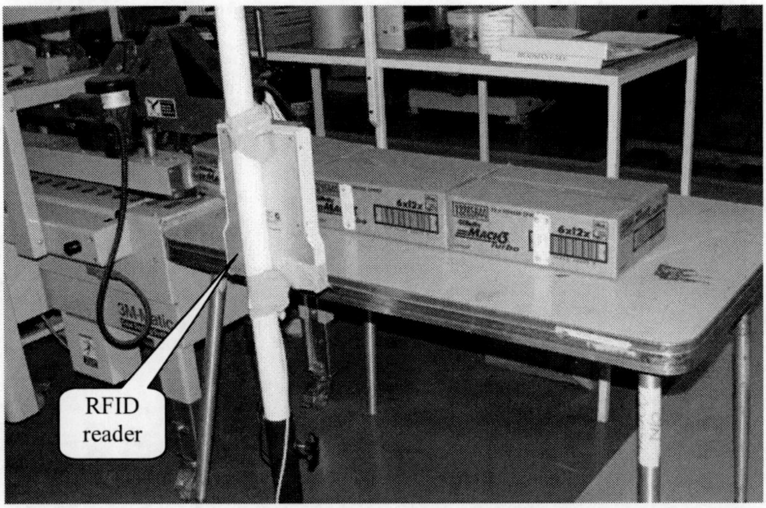

Figure 1. The "pick face" process.

Smart Shelf

The smart shelf was a shelf with RFID readability that checked the presence of tags all the time. The reader reported the tag list at regular intervals. Figure 2 shows the shelf in operation. Since each reader had 4 antenna ports, four shelves were equipped with antennas. The shelf identity was associated with the identity of the antenna attached to it.

The principle of the smart shelf was straightforward. The presence of an item was indicated by the presence of the EPC of that item. However, due to position interference, not all the tags in the space could be detected. Only the tags exposed at the top of the stack could be read. The other tags could only be read until they were exposed, i.e. after the top items were removed. Hence, the absence of an EPC would not mean the consumption of that item. The detection logic was designed such that the item should appear and then disappear before the item was declared "out" of the rack. There were other issues in this logic, for example, if the pallet was re-stacked. Nevertheless, the NDP demonstrated that this concept could work but more research would be required to streamline the process.

Figure 2. Smart shelf showing relationships of antenna, tagged items and pallets.

Wrapping and Labeling of Re-packed Shipments

When the pallets were re-packed with goods for local distribution, the items on the pallets were still loosely stacked. In order to ensure integrity of the pallet and its associated data, the pallet was wrapped with plastic sheets (Figure 3). In addition, a RFID enabled shipment label, a SSCC, was applied to the plastic. After wrapping, it was not possible to add any further item to the pallet without destroying the wrap and SSCC.

Figure 3. Shipments were wrapped with plastic sheets and a separate SSCC was added.

Search for EPC

The information gathered from these RFID processes were consolidated to the local EPC information server (EPC-IS). In order to share information securely among partners, the NDP portal was set up on a secured global server. Partners could access product data, containment (i.e. hierarchy of EPC), track and trace information. Figure 4 shows the search for the content of a purchase order using one of the EPCs as the key. This data sharing capability was found to be most useful to the supply chain partners since product information was available almost immediately after the wrapping and labeling process.

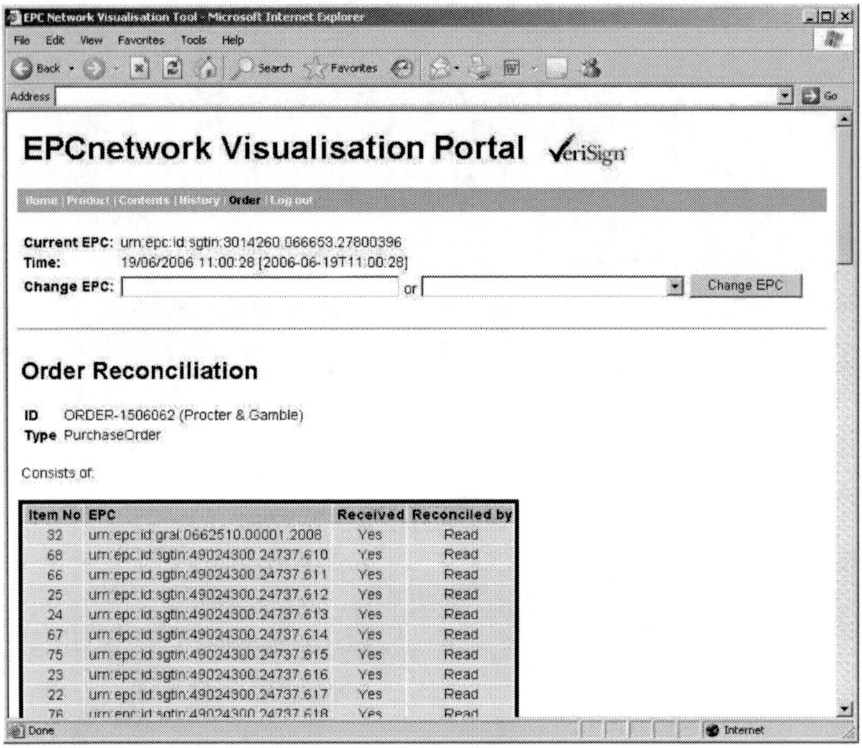

Figure 4. Track and trace for items in an order using an EPC as the search key.

THE NDP EXTENSION

The success of the NDP encouraged some of the partners to continue research on how EPC data could be used to drive business transactions. The extension project "National EPC Network Demonstration Business Information Integration" (NDP Extension) [19] concentrated on track and trace of assets, and the management of transactions on these assets with EPC.

The assets used in NDP Extension were pallets (Figure 5). The pallet industry practice was for the supply chain to hire pallets for moving goods. Hence, the pallets were owned by the pallet company and should be traced in the supply chain. However, as pallets moved among trading partners, there were always discrepancies, for example, broken or lost pallets. These irregularities were hard to detect on the spot. It was even more difficult to reconcile at a later date. The discrepancy in actual number of pallets had to be negotiated between trading partners, each of whom had their own version of events and supporting paperwork. Since there were more than ten million pallets in circulation throughout Australia, the potential for loss was enormous and the cost of managing these assets was high.

Figure 5. Tracking pallets in the NDP Extension.

In the NDP Extension, six sites, three in New South Wales and three in Victoria were installed with the EPC system. Initially test runs were unsatisfactory because the readers were unstable in detecting the RFID tags passing through their RF fields. It was later found that there were quality problems of the tags. Some tags were coded incorrectly. Some tags had manufactured problem and they were inherently weaker tags. Since the consortium used 3 different brands of readers, there were also site and reader specific issues. Each reader was tested thoroughly as a standalone unit as well as when it was connected to the EPC Network. These problems were rectified by a number of remedial measures.

The consortium developed an add-on PDA system that linked to the EPC-IS to communicate order information to the truck driver. An average efficiency gain of 24.3% was recorded for the pallet supplier, whereas an average efficiency gain of 13.9% was recorded for the pallet users [20]. The efficiency gain and hence cost reduction was mainly due to the elimination of data entry, verification and reconciliation processes. The gain was significant, especially for the pallet supplier. Furthermore, improvement in inventory accuracy as well as improvement in quality area, such as accuracy and transparency of information and real time processing had great impact on the other logistics operations such as planning and forecasting.

In addition, the project developed a business scenario in which the EPC Network could be provided to the general public as a subscribed managed service. This business model had the potential to change the expenditure on RFID infrastructure as an operating expenditure as compared to the common model of high capital investment cost.

Enterprise Model of the Two Demonstrator Projects

Both national demonstrator projects were completed successfully, but the industry was still unsure whether the technology was mature enough for global roll-out in the supply chain [21]. Using the experience in the two national projects, this chapter explores future RFID enabled supply chain models that have the potential of adopting and benefiting from new technologies.

From the enterprise modelling perspective, a supply chain is a virtual enterprise that the companies involved are collaborative and networked without formal organizational structure. There are frequent changes in products, services, processes, participating organizations, markets, logistics and distribution networks. The companies in the supply chain form a temporary alliance to deliver a project or product, but they dissolve when the task is completed. The relationship between the companies relies on trust and de facto industry practices. Success for achieving the goal therefore demands well-coordinated agility in all internal and external aspects of the virtual enterprise. However, this requires an understanding of aspects of the virtual enterprise like project coordination, resources management, team building, document management and information control were not adequately managed.

The methodology used in analyzing the two demonstrator projects is the GRAI Integrated Methodology (GIM) by Chen *et al* [22]. According to GIM, there are three systems working in a manufacturing enterprise, viz, physical, decision and information systems. These systems exist irrespective of whether it is a virtual enterprise or a single company. Figure 6 shows an adopted virtual enterprise model of the NDP.

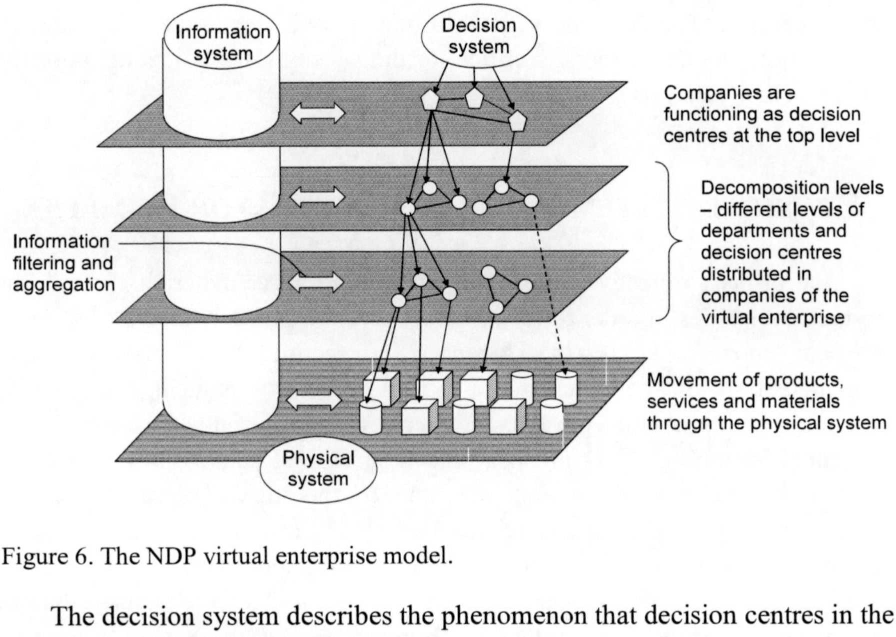

Figure 6. The NDP virtual enterprise model.

The decision system describes the phenomenon that decision centres in the enterprise are usually organized in a hierarchical structure. Decisions centres can be office bearers or autonomous units at different levels in the organization. At the top level, decisions are policy in nature whereas office bearers at lowest level make decisions affecting individual machines or products only.

The physical system describes the machines, components and resources that the enterprise has at its disposal for generating profits and wealth. The activities of the physical system are affected by the decisions made in the decision centres. In fact, all events in the physical system can be traced back to certain decisions in the decision system.

An enterprise does business through its physical system. For example, in manufacturing enterprises, decisions must drive the machines to produce. In logistics enterprises, decisions drive movement of goods through the supply chain. To facilitate the propagation of decisions to the lower levels of the decision system and the physical system, the information system plays an important role in the enterprise. If the right information about the decision made is transmitted to the right object in the physical system at the right time, the physical system will act correctly as expected by the decision maker, otherwise, the physical system will have no idea of what is expected and hence it will not take any action. In the virtual enterprise, this propagation of information is particularly difficult because of the cross company boundary characteristics. The timeliness of information and its correctness duration propagation among partners is critical to the success of the operations of the whole virtual enterprise.

Physical System in the Demonstrator Projects

The primary objective of a supply chain is to move materials through the partner companies in order to fulfill business requirements. The proof of these material movements is the business transactions that involve paper documentation and communications. In the EPC environment, the proof is captured by recording the presence of the relevant RFIDs at tactical locations.

Since capturing RFID physical data is a critical activity in the physical system, hardware configuration was one of the major issues in the two demonstrator projects. To ensure that the RFID readers can capture any tag passing through the dock door of the distribution centres, the antenna configuration in NDP was a portal configuration with two antennae mounted on steel posts at both sides of the door (Figure 7).

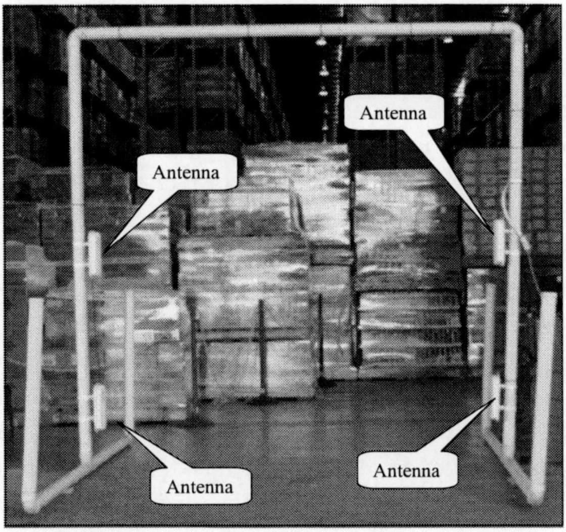

Figure 7. Dock door portal configuration with four antennae.

For the NDP Extension, due to the height of the stack of pallets, all available antennae were mounted on the readers to a one-sided configuration (Figure 8). When the pallets were moved through the dock door, they were arranged such that all tags faced the antennae side of the portal.

Figure 8. Antennae mounted as one-sided configuration in NDP Extension.

THE EPC NETWORK AS THE INFORMATION SYSTEM

The information system connects the physical system to all levels of the decision system. Hence, it is important that the information system gathers information and filters appropriately according to the needs of different decision levels. In the EPC Network, this was done by the middleware, which converted multiple reads to one read, added contextual information such as the read location, and format data for storage in an EPC data repository. The EPC filtering function was developed to comply with the EPCglobal standard.

Data transmission and presentation were managed by defining some rules and storage dealing with the data via the transmission medium. In the NDP, local information systems were configured to represent each company's product set that was tagged to provide a means of relating the numbers read by readers to a specific project participant and product. For case based transactions, entries were made to represent each of the unique Stock Keeping Units (SKUs) for the cartons to be tagged. For shipment based transactions, entries were made to identify the shipments (SSCC) that would be tagged.

The information system was divided in two levels. EPC data were captured and stored at a local server on every site and the middleware system selectively uploaded part of it to the global server. The global server therefore stored information from all local servers. The global server would also control access of information across local servers, for example, product information was held globally and the local servers had to make authenticated queries to retrieve such information for transaction processing. Detailed product delivery information was then made available to the warehouse operators who could then finalise the receiving transaction. However, completion of the purchase order still relied on the paper documentation that accompanied the goods.

In the NDP Extension, data captured in the physical system were transmitted to a central server immediately. A major difference between the two demonstrator projects in the information system was the way data integrity issue was handled. A "pick face" process was used in NDP when the goods were first packed to a pallet for shipment. There were several versions in different partner companies but all of which were designed to capture all items being shipped with 100% certainty. This information was left in the global EPC database and reconciliation of missed or error readings could be made to known items on each shipment. This process required significant inter-enterprise process control as the movement of goods (at the physical layer) and the movement of data between the decision centres were in different routes so consistency of process parameters must be maintained by management system design.

On the contrary, there was no "pick face" process in NDP extension. Capturing of RFID tag data relied on the control of a gate read process which was not 100% reliable. There was no reconciliation process for RFID read errors, but the issue was overcome by the fact that pallets were stacked uniformly and hence missed counts were corrected by the truck driver and the warehouse operators on the spot. Any further mismatch of data not picked up by the operators was accepted "as is".

A further enhancement of the NDP Extension information system was the use of Personal Digital Assistant (PDA) that integrated with a specially designed business transaction system in the pallet company server and displayed the results of RFID reads on the PDA of the trucks. This information system component enabled decision to be made on the spot.

The Decision System

In the national demonstrator projects, the decision system is a set of rules and protocols that the supply chain partners use to collaborate and transact. This represents the highest level of agreement that governs the operations within the supply chain. Companies and their departments then represent the decisions centres where decisions are made according to the protocols and agreed guidelines in the virtual enterprise. Communications between companies can occur at any level while the goods are moved in the physical system driven by the decisions in different decision centres. Decisions and supporting information are transferred to other levels via an information system, which can be a paper based system.

Decisions in the FMCG supply chains of the demonstrators were triggered by a purchase order, which started the "pick face" process. The data generated in this decision were a structure of item EPCs (also known as Serialized Global Trade Identification Number, SGTIN), SSCC and the pallet EPC (also known as Global Returnable Asset Identifier, GRAI). Once these data were created, the pallet of goods could be loaded onto trucks.

In the NDP Extension, since there were assets only, SGTIN and SSCC were not required, and there was no decision centre on shipment. However, the consortium examined the decision process for acknowledging receipt of pallets and designed a new process supported by PDA.

Under the new process, when a forklift load of pallets was driven past the fixed readers at the pallet service centre, the tag information was delivered to the central server to process the tags. As the central server had no reference information for the order and therefore could not tell if the correct number of pallets was read, the tag data and counts were sent to the business transaction system via a web services message. The business transaction system paired the tag reads to the original order information and synchronized the tag count to the truck driver's PDA where a traffic light display indicated whether the order had been completely loaded. There was also a count so that the driver could see precisely how many pallets had been read and loaded. Decision on accepting the transaction could then be made on the indicator information. The two decision processes are shown in Figure 9.

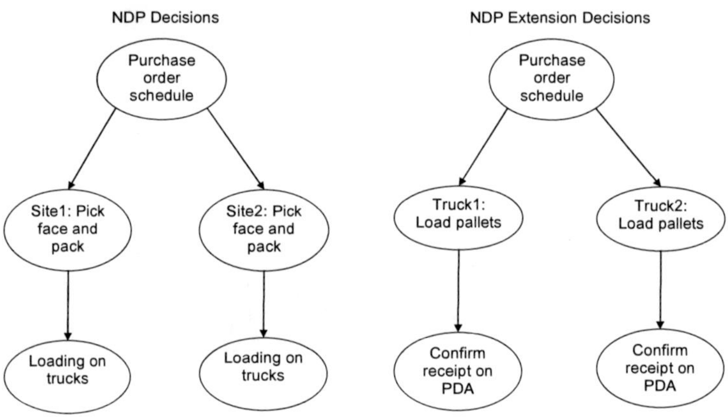

Figure 9. The demonstrator projects decision processes.

FUTURE VIRTUAL ENTERPRISE MODEL FOR RFID ENABLED SUPPLY CHAIN

To formulate a future virtual enterprise model, lets imagine a manufacturer of finished goods in China receiving order information via a messaging standard. The goods are then packed into correct configuration from the information obtained from the electronic order. Once the products are packed, the information is sent back to a central RFID repository where the information such as location, status of a particular tag are updated in real-time.

An international forwarding agent is then notified and makes necessary arrangement to ship the freight to the correct destination. Tag information from the central RFID data repository can also be accessed by customs agency around the world, hence reducing clearance time. Once the goods are on-shore, they are unpacked for further distribution. Since each package is already assigned a specified distribution centre or retailer outlet, the products are simply cross-docked rather than received into inventory. Several important characteristics are identified in this scenario:

1. EPC technology is designed on the premise that the same RFID infrastructure is used throughout the supply chain. However, if one of the participants is not EPC compliant, for example, using Japan's UIC [23] or China's NPC [24], the information link for the entire supply chain cannot be assured.
2. Partners of the supply chain change frequently, as it is understood to behave as a virtual enterprise. The virtual enterprise system should be as easy for entry and exit as possible. However, existing RFID infrastructure is expensive due to compliance restrictions on system configuration, for example, fixed IP address is required for the EPC-IS and readers. The companies in the national demonstrator projects hesitated to roll out the technology to other parts of the enterprise.
3. It has been observed in the national demonstrator projects that RFID tags are not 100% reliable. The virtual enterprise system should have fall back position or contingency plan to deal with non-performing tags.
4. The NDP Extension used a mix of brands of RFID readers. Interoperability between these readers was an issue.

A new virtual enterprise model is required for a RFID enabled supply chain to understand the minimum adaptation requirements. More efficient implementation and operations will result in reduced cost of inventory for supplier and distributors, reduced cost of handling, increased efficiency and productivity. The model in Figure 10 uses the extension of the information system beyond the EPC Network structure. The IT manager on site has the responsibility of managing the RFID configuration deployed to reading locations. This part of the architecture still needs further work among the RFID vendors to make it a plug and play business environment. Extensions to new non-EPC data sources are allowed as add-on. Using a similar service aggregation concept [25], data from all sources in the information system will be pre-conditioned by the unified filtering and aggregation service prior to presentation to the decision system.

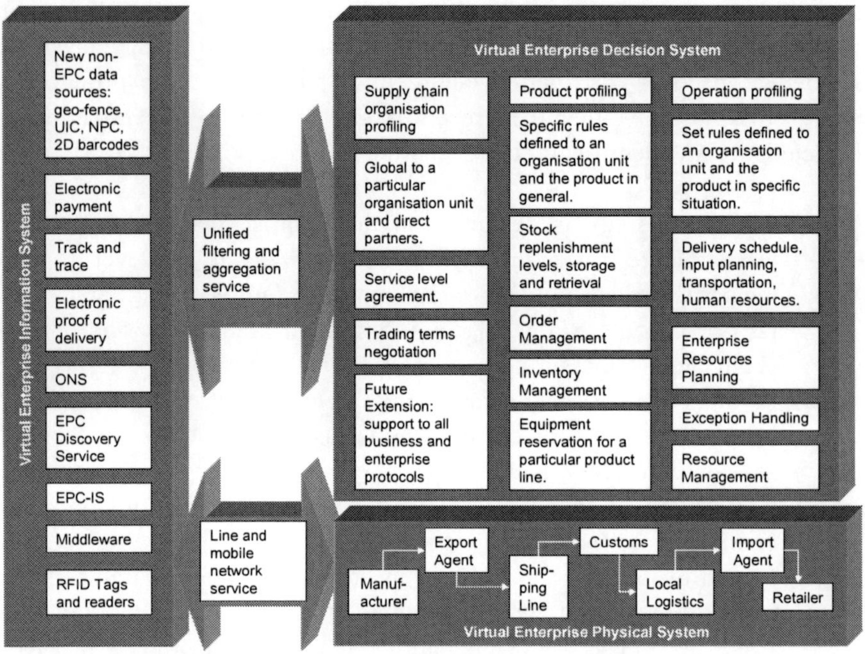

Figure 10. System architecture for the heterogeneous RFID enabled virtual enterprise.

In the physical system, the focus is not on movement of goods, but rather on the operators on the goods. These operators will adopt the system's principle of detect-and-read-and-record. Data transmission to the information system will be implemented via new pervasive networks such as wireless. The reliability issue of RFID tags can be subsumed by the abundance of data not only from RFID sources but also from non-EPC data sources. Availability of massive data in the information system will enhance the ability of the system to cross-validate or triangulate-verify product and shipment data. This will increase the chance of reconciling any data discrepancy significantly.

It is also necessary to note that the RFID enable virtual enterprise should adopt the universal business rule for packaging. Once the goods are packed in the "correct configuration from the information obtained from the electronic order", the package is unpacked only when it arrives at the distributor's centre and unpacking RFID process is taken place. Hence, so long as the packing information is secure, the package identity will remain unchanged until the other end.

The decision system should be flexible enough to accommodate new rules and practices due to addition or departure of companies. This is particularly important for SMEs since they join the virtual enterprise for a reason and will leave as soon as the system is unfavourable to them. The concept of open access infrastructure is therefore key to this implementation. The business functions of individual enterprises are publicized in web-portals as rules and protocols within the global supply chain environment. Different organization types and objectives can then be accommodated, with consideration for interfacing among organizations through a virtual interface.

EXAMPLE OF NON-EPC DATA SOURCE: GEO-FENCE

The issue of high infrastructure cost for RFID implementation has attracted serious questions in relation to its suitability in the supply chain [26]. Alternative technologies are being investigated, one of which is global positioning system (GPS). Since late 1990's, it has been a common trend for 3PL companies to incorporate GPS technologies via mobile data network such as GSM or GPRS to track their vehicles and driver [27]. If the data source is integrated with specific business functions, logistic services can be billed on an hourly basis [28].

Existing RFID tracking frameworks are based upon physical RFID tags being read by RFID readers which reside over a gate or doorway. As products containing RFID tags move across the RFID readers, the location of the products is confirmed. This feature may be very useful for traceability in a warehouse or a library, but it may not be necessary when products move across two trading parties.

To extend RFID functionality beyond the warehouse boundary, the PDA (as described in the NDP Extension case study) can be configured as a mobile server linking to GPS data. A virtual boundary on the GPS map called "geo-fence" is defined (Figure 11). The mobile server keeps track of the GPS coordinates against the defined geo-fence. When a vehicle enters the valid geo-fence zone, an arrival timestamp is stored. An event will be triggered and a set of predefined tasks, such as update job status or even update on an EPC-IS database, will be actioned. Similarly, the departure timestamp can be stored and the vehicle loading time can be calculated.

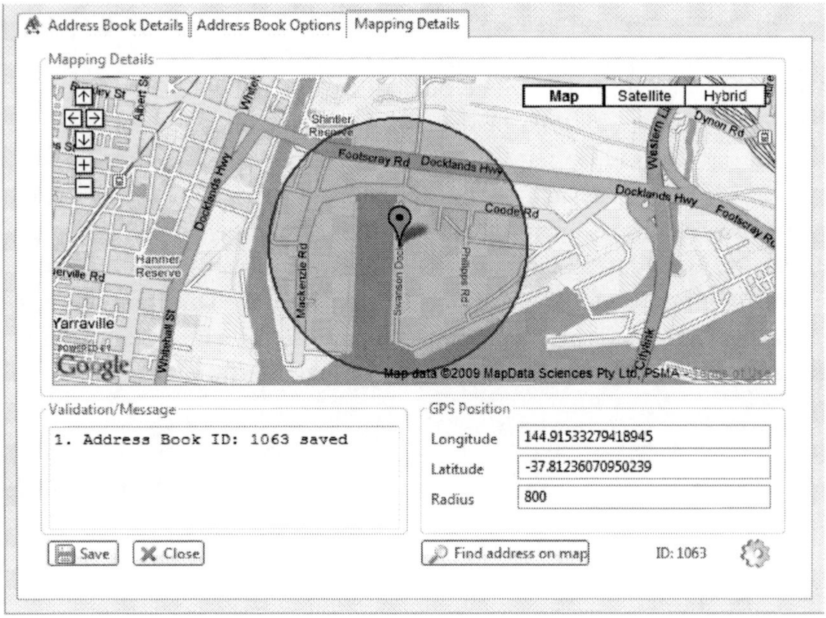

Figure 11. Circular geo-fence detection process.

This extended RFID track and trace capability can be taken further to incorporate it at item level. To achieve this goal, the geo-fence capability needs to be enhanced with an identification mechanism. The concept is to link the geo-fence data with RFID data including shipment (SSCC) and asset identification (GRAI). This RFID traceability capability works well within the company's boundary. The event management of the local system will update the consignment with all EPCs of goods and the truck that is assigned to this route. Once the truck leaves the depot, the connection to EPC Network of the consignment will be transferred to the geo-fence environment.

Making use of the geo-fence system, the extended infrastructure requires integration of several key functions, using a "universal" system to manage the processes. For ease of explanation, the "universal" system is coded name "Transparent" in this chapter. Transparent functions as the unified filtering and aggregation service that allows non-conforming systems to interact with other conforming system such as those of EPCglobal. However, the primary goals should not only be constrained to traceability and visibility. Instead, management of information should take higher priority so that decision paths could be established. Information gathered across the framework could then be synchronised with shipping line, terminal operators, customs agencies, empty container parks or even local traffic to further optimise the performance and the transparency of each process within the entire supply chain. Figure 12 shows the overall design of the extended infrastructure.

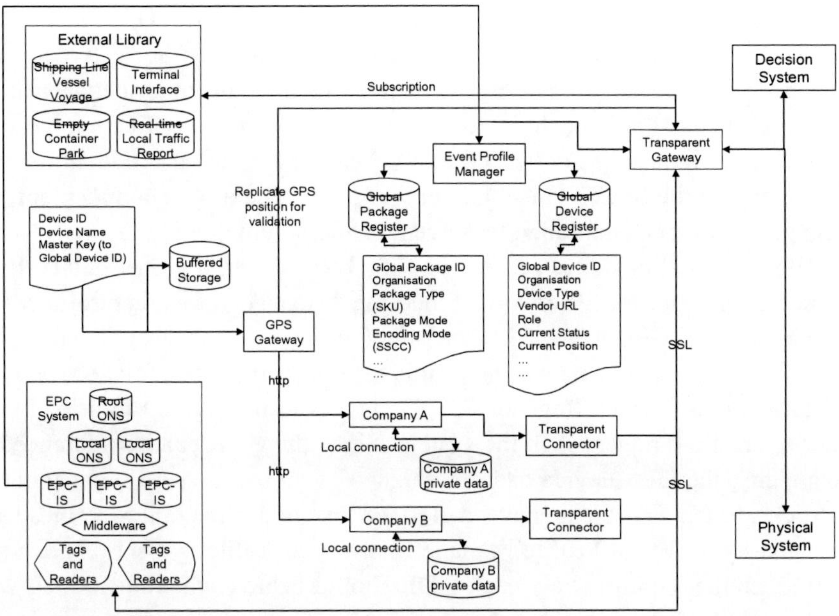

Figure 12. Overall design of the unified filtering and aggregation service.

The focus in Figure 12 is to keep track of each SKU. When ASNs are received, the warehouse needs to accurately account for each item inside the package, and put them away to their set location inside the warehouse.

The supply chain process is reversed when an order is issued to the warehouse to deliver goods to another party. The SKU from a defined location is packed onto a pallet with a SSCC attached. An ASN is then sent to the recipients with a list of SSCC and items contained inside each SSCC. It is at this stage that a freight label is created and then applied to the SSCC. The freight label does not have any correlation to the SSCC. All it does is simply tell the truck driver that how many packages he is picking up per consignment. It is in this transition that the link is broken, and in most case two separate systems are used to capture this information. When the freight arrives at the recipient, only the package details are capture, not the SSCC. The recipient will then need to re-identify with the SSCC by checking or scanning the actual package.

The problem becomes more complex when the barcodes are substituted by RFID tags. Since the warehouse and the transport company may not be the same, two different tags may be used for the same purpose, which adds more costs to the system and defeats the integrity of the identity. Security concerns have been raised that the package will need to be kept intact between dock locations. Typical approach is to enhance the normal verification process known as "pick face" (as described in NDP case study) to ensure identification of the items to the SSCC. Once the items are validated as one containment, the package is wrapped and sealed to prevent further alternations.

When a local device is registered with Transparent, a global device ID is created within the local network. Since the extended infrastructure needs to maintain open access by any SMEs wishing to join the global supply chain, the local device ID cannot be assumed unique. Encapsulation of all local device IDs in a Global Device Register (GDR) class is to manage ID information in order to ensure uniqueness in the system. Since the GDR can be referenced to a company, the GPS device can be assigned as global device ID dynamically to trucks. Each GDR can have a defined event profile build-in, these are managed by an Event Profile Manager. An event profile is a set of rules and parameters that controls how each GDR should behave when interacting with another GDR at a specified condition.

The same principle applies to the Global Package Register (GPR). Since EPC network is not the only network in the global supply chain, the system will need to encapsulate each tags in a GPR class. For example, RFID information can flow to and from different networks, such as NPC, UID as well as within the virtual network.

Since the SSCC as well as the physical RFID tags number are embedded inside the global ID registers, the same set of information can be filtered across to the physical network. Also the status and event of the physical system can also be transposed to the virtual system. Data travel between physical networks is managed by Transparent Gateway.

Data from the geo-fence are fed directly to the local company's database to trigger an event when a vehicle travels within the geo-fence at a defined moment in time. Since this GPS data are critical in terms of creating an event, such as changing ownership, etc, it is not desirable to depend on one source of data. The problem is that since the data are kept within the company's private database, it can easily be tampered with and therefore would be considered unreliable. One solution is to allow the same set of data to also pass in to the

Transparent Gateway. When an event is triggered from a local database, it is verified by the data that are collected directly from the buffered storage. Since the geo-fence are independent and cannot be tampered with, if both set of data are identical, the event triggers are valid, otherwise, the relevant parties are notified and security investigations will be initiated.

SUPPORTING GLOBAL OPERATIONS

The issue of global operations is not so much on gathering of data, but is about interpreting them in a meaningful way that drives business requirements. For managing sea freight, information from various shipping lines need to be collected and synchronized before a freight operator can make a judgement on when and where to collect a particular sea freight containers. Ships that carry containers from port to port are often referred to as a vessel. A Lloyds number is a unique number that identifies each vessel, similar to a license plate of a car. A Voyage number is related to each trip. Therefore a ship has an in-voyage and an out-voyage every time it arrives at a port. An in-voyage represent import container where an out-voyage represent export containers. In some cases, it is possible for a ship arriving at one port, where container are unloaded and leave for another terminal without taking any load.

Before a ship arrives at a port, each port operators (terminals) must publish a shipping schedule which is then sent to the registered carriers to pickup containers from the terminal. This information is also available at each terminal web portal, as well some of the shipping line website (Table 1).

However, the web portal does not have enough container shipment information to tell what is in that consignment (for the purpose of further distribution when the container arrives). Some data could be wrong, for example, the vessels are delayed. The terminal will re-publish the updated shipping schedule, but the update only occurs locally at the web portal, therefore each carrier will need to re-check each terminal's shipping schedule multiple times per day to enable them to make decision of when and where to pickup the container.

To ensure that each vessel is loaded with as little delay as possible, export containers are received before import containers are unloaded. Before a container can be collected or delivered to each terminal, a carrier will need to request a time-slot (when and what container are to be collected) from each terminal. Carriers are penalized when they run late or do not turn up for their time slot.

Table 1. Information available from shipping web site

Name	Description
Vessel Name	Unique Name of the vessel
Lloyds Number	Unique Number of the vessel
ETA	Estimate time of arrival (for import containers)
In Voyage	Inbound voyage number (for import containers)
Available From	First day which carrier can time-slot and pickup the containers
Available To	Last day which carrier can time-slot and pickup the containers before storage charge commence
ETD	Estimate time of departure (for export containers)
Out Voyage	Outbound voyage number (for export containers)
Receiving From	First day which carrier can time-slot and deliver the containers to the terminal
Receiving To	Cut off date which carrier can time-slot and deliver the container to the terminal

Before a container can be collected, it must be cleared by Customs. This will ensure that the duty of the commodity inside the container is fully paid before collection. A container can also be collected if a request to re-locate the container to a bond store, which is a holding facility where container are unpacked and commodity are stored. Items inside the container can be picked up partially or in full once all duties and taxes are paid. This is common for most high value items.

Once the import container has been time-slotted, a carrier will often check each terminal's web portal to see if the cargo has been custom cleared to pick up. Usually the custom agent holds on the duty until the last day before the container is picked up. To facilitate these interests, a communication path exists between Customs, the shipping line, terminal operators.

A Quarantine and Inspection Service officer will check the container before they leave the terminal's exit gate. If a container loaded on the trailer contains dirt, it will need to be re-directed to steam-clean. However, if the commodity inside container contains wood, it will have to be fumigated. This is an unplanned event which will ultimately delay the delivery of the container. Having the extended RFID infrastructure in place, the content inside each container is known in advance before it arrives. The quarantine event can then be planned accordingly.

Figure 13 shows the information flow between various terminals, shipping lines and Customs. Using Transparent Gateway, all the public (vessel and voyage details) are synchronized and private data (container status) for each carrier. The data feed back from private Transparent Gateway can also trigger event which can update container status inside the each company's enterprise resources planning (ERP) system.

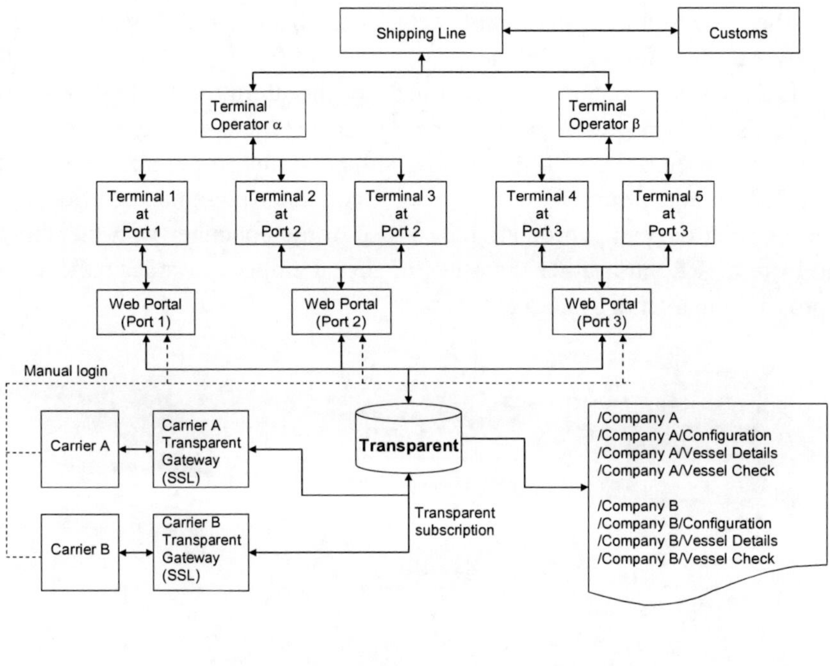

Figure 13. Information flow between various terminal, shipping line and Customs.

Detention charges occur when a vehicle is detained outside the agreed parameters. This usually involves the consignee taking too long to load/unload the containers. However, this charge is usually applied after the detention would occur, which can lead to payment dispute, since the freight forwarding may not have enough time to invoice the consignee. However, with the use of geo-fence, each truck and its items are monitored continuously. As soon as the truck enters a geo-fence of the consignee, the time is locked. If the allowable detention time is reached, an alert such as an email or SMS could be sent to the freight forwarder or directly to the consignee informing them that detention is about to commence. If there is any dispute, the matter can be discussed before invoicing takes place. When the vehicle leaves the geo-fence, the duration time calculated. If it is greater than the agreed allowable time, the detention time charge is added to the current consignments.

In most cases, wharf cartage will also involve de-hiring of empty container for an import operation. A carrier does not have a choice about where the container will be de-hired. Very often, the empty container park may be congested due to their set operating hours. Therefore if one or more vehicles are known to be congested, then an alert could be sent to the fleet controller to re-direct the empty container back to the yard, to avoid congestion and improve efficiency.

If an empty container is returned to the third party logistics (3PL) company's container yard, it may be possible for the company to publish the details of the empty container details (shipping line, type, size, etc) so that other 3PL companies nearby, who need a container of same specification for export operation, can coordinate the release of the container with the shipping line from a 3PL yard rather than the congested empty container park, thereby improving the overall system efficiency.

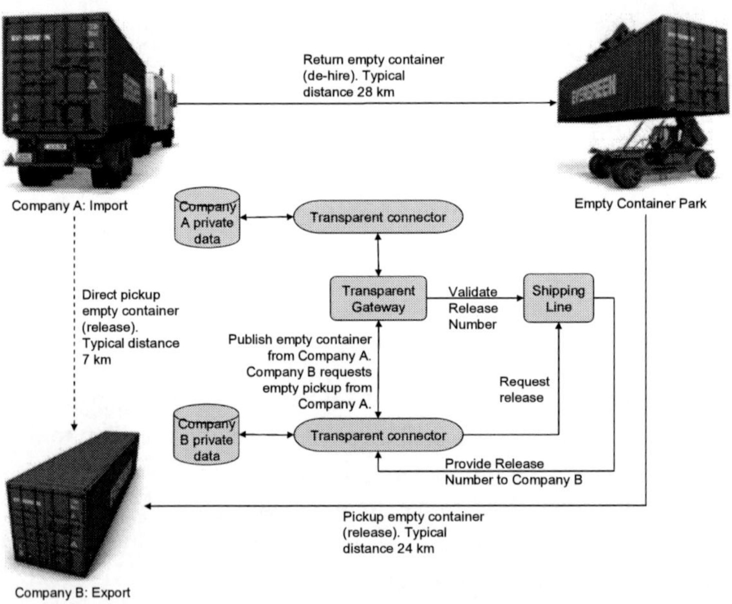

Figure 14. Redirection of empty containers.

CONCLUSION

From the national demonstrator projects, it is clear that there are advantages and disadvantages of using RFID for information capturing and inventory management in supply chain. The advantages of RFID have been identified including no need for line of sight, co-use with barcode and RFID tag, and data transparency using EPC standard. RFID can be best implemented in a supply chain for tracking movements of materials within manufacturing environment and between trading partners in the supply chain. In many cases, RFID systems can be used to monitor unpacking and re-delivering freight into a distribution centres or retail outlets.

However, the biggest stumbling block in any RFID application is that the system must deal with incomplete data due to reliability of the RFID system and non-compliance of some parts of the global supply chain. To ensure a complete analysis and fool proof design of the system, enterprise modelling methods have been used. Enterprise models can assist to examine elements in the system and develop a new model from which enterprise execution system can be developed.

In this chapter, a virtual enterprise model for RFID enabled supply chain has been developed from the experience of the national demonstrator projects. A new virtual enterprise model pointing to an extended infrastructure is introduced. To support this new model, the concept of geo-fence is described as a potential non-EPC data source. A geo-fence tracking device is made possible by integrating GPS system with mobile devices. The system is enhanced with a unified filtering and aggregation system Transparent that manages a variety of devices in a global supply chain. By properly designing the underlying global registers and connectors, the extended infrastructure works with existing RFID frameworks as well as non-EPC compliant systems. The model allows alternative technologies to verify the position of the RFID tags and then update the existing RFID framework to ensure that the traceability of the RFID tags remain intact. To illustrate its use, a container management scenario is described using Transparent to synchronize with web-based information from the shipping companies.

REFERENCES

[1] Pawar, K.S., Driva, H. (2000). Electronic trading in the supply chain: a holistic implementation framework. *Logistics Information Management.* 13(1):21–32.

[2] Myer Pty Ltd. (2008). Supply Chain – Merchandise Logistics E-commerce overview guide. October. 7 pages. Available from: http://myersupplier.myer.com.au/documents/E-Commerce%20Overview%20Guide.pdf

[3] Power, D.J., Sohal, A.S. (2002). Implementation and usage of electronic commerce in managing the supply chain - A comparative study of ten Australian companies. *Benchmarking: An International Journal*, 9(2):190–208.

[4] Angeles, R. (2005). RFID Technologies: Supply-chain applications and implementation issues. *Information Systems Management*, Winter:51-65

[5] Xiao, Y., Yu, S.H., Wu, K., Ni, Q, Janecek, C., Nordstad, J. (2007). Radio frequency identification: technologies, applications, and research issues. *Wireless Communications and Mobile Computing*. May, 7(4):457–472.

[6] Sellitto, C., Burgess, S., Hawking, P. (2007). Information quality attributes associated with RFID-derived benefits in the retail supply chain. *International Journal of Retail & Distribution Management*, 35(1):69–87.

[7] Ranky, P.G. (2006). An introduction to radio frequency identification (RFID) methods and solutions. *Assembly Automation*, 26(1):28-33.

[8] Visich, J.K., Li, S., Khumawala, B.M., Reyes, P.M. (2009). Empirical evidence of RFID impacts on supply chain performance, *International Journal of Operations and Production Management*, 29(12):1290-1315.

[9] Vijayaraman, B.S., Osyk, B.A. (2006). An empirical study of RFID implementation in the warehousing industry. *The International Journal of Logistics Management*, 17(1):6–20.

[10] Deitz, G., Hansen, J., Richey, Jr.R.G. (2009). Coerced integration: The effect of retailer supply chain technology mandates on supplier stock returns. *International Journal of Physical Distribution and Logistics Management.* 39(10):814–825.

[11] GS1 Australia, CSIRO (2006). EPC Network™ Australia Demonstrator Project Report. Available on request from http://www.gs1au.org/.

[12] GS1 Australia, RMIT University (2007). National EPC NetworkTM Demonstrator Project Extension. Public report distributed in Smart Conference 2007, 20–21 June, Sydney. Also available on request from http://www.gs1au.org/products/epcglobal/australian_activities/demonstrator_project.asp.

[13] Soppera, A., Farr, J., Kasten, O., Illic, A., Zanetti, Z., Harrison, M. (2007). Supply Chain Integrity (D4.6.1), Building Radio frequency IDentification for the Global Environment (BRIDGE Project Report), Dec, WP04, European Commission contract No: IST-2005-033546, 22 pg..

[14] Lawson, R. (2002). The implementation and impact of operations strategies in fast-moving supply systems. *Supply Chain Management: An International Journal*, 7(3):146–163.

[15] Daverport, T.H., Brooks, J.D. (2004). Enterprise systems and the supply chain. *Journal of Enterprise Information Management*, 17(1), 8–19.

[16] Samaranayake, P. (2005). A conceptual framework for supply chain management: a structural integration. *Supply Chain Management: An International Journal*, 10(1), 47–59.

[17] IFIP–IFAC Task Force on Architectures for Enterprise Integration, (1999). *GERAM: Generalised Enterprise Reference Architecture and Methodology Version 1.6.3, Annex to ISO WD15704, Requirements for enterprise-reference architectures and methodologies*. March, pub: IFIP and IFAC.

[18] Mo, J.P.T. (2008). Development of a National EPC Network for the tracking of fast moving consumer goods. *International Journal of Enterprise Network Management*, 2(1):25–46.

[19] Mo, J.P.T., Gajzer, S., Fane, M., Wind, G., Snioch, T., Larnach, K., Seitam, D., Saito, H., Brown, S., Wilson, F., Lerias, G. (2009). Process integration for paperless delivery using EPC compliance technology, *Journal of Manufacturing Technology Management*, Vol.20, Iss.6, pp.866-886.

[20] Gajzer, S. (2007). Process Modelling and Development of the Key Performance Indicators for the National Electronic Product Code Network Demonstrator. MEng Dissertation. RMIT University.

[21] Wamba, S.F., Keating, B., Coltman, T., Michael, K. (2009). RFID Adoption Issues: Analysis of Organization Benefits and Risks. Pub: Centre of Business Service Science, University of Wollongong, April, ISBN: 978-1-74128-170-5, 16 pages.

[22] Chen D., Vallespir B., Doumeingts G., (1997). "GRAI Integrated Methodology and its mapping onto generic enterprise reference architecture and methodology", *Computers in Industry*, Vol.33, pp.387–394.

[23] Ubiquitous ID Centre (2006). Ubiquitous ID Architecture. *T-Engine Forum*. Document 910-S002-0.00.24 / UID-CO00002-0.00.24.

[24] Brown, D.E. (2007). RFID Implementation. Pub. McGraw Hill. ISBN-13: 978-0-07-226324-4 and ISBN-10: 0-07-226324-5.

[25] Piccinelli, G., Vitantonio, G.D., Mokrushin L. (2001). Dynamic service aggregation in electronic marketplaces. *Computer Networks*. Elsevier, 37(2):95–109.

[26] Reyes, P.M. (2007). Is RFID right for your organization or application? *Management Research News*, 30(8):570–580.

[27] Zito, R., D'Este, G., Taylor, M.A.P. (1995). Global positioning systems in the time domain: How useful a tool for intelligent vehicle-highway systems? *Transportation Research Part C*, 3(4):193–209.

[28] Hamilton, J.W. (1993). Wireless communication systems: A satellite-based communications approach for competitive advantage in logistic and transportation support services. *Computers in Industry*, April, 21(3):273–278.

Reviewed by Dr. Laszlo Nemes.

In: Radio Frequency Identification ISBN 978-1-61122-416-0
Editor: Alison R. McAdams, pp. 31-53

Chapter 2

THE ROLE OF RFID IN AGRICULTURE

Luis Ruiz-Garcia*

Physical Properties and Advanced Technology in Agrofood.
Universidad Politécnica de Madrid, Spain
ETSI Agrónomos. Edificio Motores. Avda.
Complutense s/n 28040 Madrid, Spain

ABSTRACT

RFID is entering in a new phase. RFID technologies are said to improve the performance of many agricultural processes. Recent advances offer vast opportunities for research, development and innovation in Agriculture. This is the consequence of lowering costs of ownership, engineering increasingly smaller sensing devices and the achievements in radio frequency technology and digital circuits. The aim of this chapter is to give readers a comprehensive view of the possibilities of RFID application in agriculture.

RFID was originally developed for short-range product identification, typically covering the 2 mm - 2 m read range and has been successfully applied to food logistics and supply chain management processes. With an increasing demand for security and safety, complete documentations for food products, from field to customer, have become increasingly demanding. RFID has been accepted as a new technology for

* E-mail: luis.ruiz@upm.es; www.lpftag.upm.es

a well-structured traceability system on data collecting, and human, animal and product tracking. Also, electronic identification of cattle using RFID is a common practice in many farms. These devices are used to monitor animal behavior in mid-size outdoor pens, providing digital data that can be easily computerized. However, recent developments in RFID hardware outfitted with sensors extend its range of application. There are commercial active and semi-passive tags that can collect temperature information. Other semi-passive tags outfitted with sensor are under development, like humidity, shock/vibration, light, pH and concentration of gases, such as acetaldehyde or ethylene. Moreover, the last generation of Class 4 RFID tags can be configured in a mesh network. In this type of network, the tags can communicate each other to get to a reader circumventing environmental obstacles and extend the size of the system.

These specialized RFID monitoring devices promise to revolutionize a wide range of agricultural operations. The development of applications in precision agriculture, like monitoring crops, makes possible to increase efficiencies, productivity and profitability while minimizing unintended impacts on wildlife and the environment, in many agricultural production systems. Instead of take decisions based in some hypothetical average condition, which may not exist anywhere in the reality, a precision farming approach recognizes differences and adjusts management actions accordingly.

In cold chain of perishable food products RFID monitoring provides new features that have the potential to be an economically viable, alerting if the products are not stored at the right temperature and predict the remaining shelf life. In this field, several applications for monitoring cold chain logistics by means of RFID have been reported.

1. Introduction

RFID is entering in a new phase ad is said to improve the performance of many agricultural processes. Although the concept of RFID is not new, last developments make possible a wide variety of applications, driven by the increasing maturity and adoption of standards, such as ISO 15693, ISO 18000 and ISO 11784[1-7]. Recent advances offer vast opportunities for research, development and innovation in Agriculture. This is the consequence of lowering costs of ownership, engineering increasingly smaller sensing devices and the achievements in radio frequency technology and digital circuits.

With an increasing demand for security and safety, complete documentations for food products, from field to customer, have become increasingly demanding. RFID has been accepted as a new technology for a well-structured traceability system on data collecting, and human, animal and product tracking. If trends continue such that the average grower has the responsibility for more land area, these advantages become more apparent. In cold chain of perishable food products RFID monitoring provides new features that have the potential to be an economically viable, alerting if the products are not stored at the right temperature and predict the remaining shelf life. Also, RFID applications are very promising in precision agriculture, in fields like environmental monitoring, irrigation, greenhouse and farm machinery. RFID helps users achieve farm management processes optimization, derive additional benefits and maximize return on investment for the farmer [8].

Both passive and active have are being used in Agriculture. Active tags are very interesting, especially for animal behavior studies. They automatically send impulses, so the animals can be identified by even distant readers. This ability is guaranteed by using a power battery. These devices can be used to monitor animals in mid-size outdoor pens, providing digital data that can be easily computerized [9].

The first applications were developed just for identification; however, growing interest in the many possible applications has led to the development of a new range RFID devices outfitted with sensors that extend the range of application. Normally, these sensor nodes consist of three components: sensing, processing and communicating [10]. Current temperature monitoring systems like strip chart recorders or temperature dataloggers are usually expensive and not automated, thus requiring manual inspection. RFID devices are more accurate and can be read without the need of visual contact [11]. There are commercial active and semi-passive tags that can collect temperature information [12, 13]. Other semi-passive tags outfitted with sensor are under development, like humidity [11, 14], shock/vibration [15], light [11, 16], pH [17] and concentration of gases, such as acetaldehyde or ethylene [18]. Biosensor tags are also been investigated. These tags could be used for detecting bacterial contamination on food products along the supply chain [19].

The aim of this chapter is to review the numerous applications that utilize RFID in agriculture and food industry and to classify them in appropriate categories. The analysis of their characteristics and contributions could be useful for perceiving new applications or relevant research opportunities.

2. RFID Applications in Agriculture

This section presents most relevant applications in agriculture and food industry. The development of these applications in agro-food has attracted considerable research efforts in the last years, because these technologies are very suitable for distributed data collecting and monitoring in tough environments such as greenhouses, cropland, warehouses or refrigerated trucks. However, some areas have been developed faster than others. For example, there are several applications in livestock or cold chain monitoring and just a few in farm machinery.

2.1. Food Traceability

Traceability is a main requirement in agro-food and has become an important issue. Traceability facilitates the withdrawal of foods and enables for all the operators of the supply chain to be provided with targeted and accurate information concerning implicated products [20]. The chain from harvest to consumer's plate forms a single entity. Tracing all components of food offered from "farm to fork", including accurate real time data, allows minimizing food safety risks, achieving a fast and effective response to incidents and increasing confidence in food products [21-23].

Policy Requirements

Global food safety policies have been stipulated by Governmental authorities and a new series of regulations were created and adopted all over the World, with particular incidence in the EU (European Union) and the United States of America (USA), as a consequence of several food incidents and scandals.

According to the actual regulation in the European Union (applicable since 2005), traceability is required in all stages of the supply chain, covering all food and feed as well as business operators without prejudice to existing legislation on specific sectors such as beef, fish, GMOs (genetically modified organisms), etc. As in the same way as the aforementioned EU regulation, in the USA, the Bioterrorism Act 2002 calls for one-up/one-down traceability for each link in the supply chain. This regulation requires that each company in the supply chain keeps information about the company that they received the products from, the company who delivered the product to them, the company who took it away, and the company they gave the products to.

In this framework, RFID has been successfully applied to traceability control in agro-food logistics and supply chain management processes because of its ability to identify, categorize, and manage the flow of goods [24-28]. The lowering cost of RFID will provide the opportunity to track and trace not only large and expensive products, but small and cheap ones, creating a new generation of intelligence products [29]. From the raw material to the sale of goods, more and more information can be gathered and made available. Moreover, this information can be linked with a traceability system in each step of the life of the product, tracking and tracing products from the field to industry in a new exhaustive way. Also, the concept of "cold traceability" has been introduced to trace groups of temperature-sensitive products are transported in different atmosphere requirements [30].

The link between RFID sensing devices and decision support systems revolutionizes refrigerated food logistics and the inventory management. If a direct access to the means of transport is not possible, online notifications offer new opportunities for improve transport planning. If fixed delivery commitments require ordering of a replacement, the time of information is very crucial. Improved cool chain management methods such as the Quality Oriented Tracking and Tracing Systems (QTT) offer new features [31]. An example of this approach is the Safety Monitoring and Assurance System (SMAS) that was developed to reduce customers' risk of consuming microbiologically contaminated meat [32]. The growth rate of pathogens was estimated based on temperature history. At a control point the package was either sent to the local or the export market. A case study of cooked ham was carried out based on previous surveys of distribution chain conditions. Following the SMAS approach, the number of products with zero shelf life could be reduced from 12 % to 4% in the export store compared to normal FIFO (First In, First Out) handling. A retailer that knew which of the cases had the shorter shelf life could put it out before the one with the longer shelf life. This is known as FEFO (First Expire, First Out) [33].

Another possibility of improvement the traceability in the food supply chains is the integration of RFID in packaging. Embedding an RFID inlay within the structure of a package, corrugated case or folding carton allow the development of "smart packaging" [34]. However, the way that RFID interacts with packaging materials is complex and with a negative impact in readability of RFID tags and labels. Readability test show that RFID would be affected by aluminum, cardboard, glass, and even stretch and bubble wrap. Also, if a corrugated cardboard gets wet, the read rates for tags on or in the boxes fall more than 50 percent [35].

2.2. Animal Identification and Tracking

Modern animal production has changed in recent years due to the use of precision tools. Results of recent research have been used as inputs to preventive diagnostics and development of decision-making software in several areas, as well as to predict events. This section provides information on current RFID animal tracking technology, how they work, current applications, and possible future direction.

In recent years a number of outbreaks of animal disease (e.g. BSE, foot and mouth disease) have revealed weaknesses in systems used to keep track of livestock and livestock products.

On average, each of the 96 million cows in America change ownership three and a half times over their lifetime, ending with the meat packer [36]. In the European Union millions of animals are imported and exported every year and transnational animal production systems are very common. In a global environment like this, awareness, fear and recognition of animal borne diseases (e.g. BSE, foot and mouth disease) have fuelled the development and implementation of reliable and effective systems for individual identification and tracking of livestock. As a result, electronic identification of cattle using RFID is actually a common practice in many farms [37]. Due to the fast movement of animals and animal food products, it is essential for the administration to be able to have "rapid trace-back" capability. Permanent identification also helps the farmer to manage more animals to be cost effective, provides better proof of ownership to reduce stock theft and improves transparency maintaining consumer confidence in animal products [8].

There are three main standards involved in animal identification with RFID, defined by the International Organization for Standardization (ISO):

- ISO 11784 describes the structure and the information content of the codes stored in the transponder [1].
- ISO 11785 specifies how a transponder is activated and how the information stored is transferred to a transceiver and it is applicable in connection with ISO 11784 [2].

- ISO 14223 specifies the air interface between the transceiver and the advanced transponder used in the radiofrequency identification of animals under the condition of full upward compatibility according to ISO 11784 and ISO 11785 [3].

Traditional forms of animal identification are considered inferior in comparison to RFID technology, while the application of RFID identification to improve farm management practices is also touched upon [8]. Traditional ear tags are lost 5 to 60% of the time, while brands or tattoos on cattle can be damaged or fade away. Traditional systems require visual detection and must be recorded manually, which can easily introduce human errors, while the labor cost of such a practice is also high. Reading errors are estimated to occur in six of every 100 animals processed via traditional mechanisms, while RFID devices are estimated to produce only one error for every 1000 animals [8]. There are four basic ways for attaching RFID transponders to the animals: collar transponder, ear tags, injecting tiny glass transponders under the animal's skin, or via a 'bolus' where the RFID transponder is mounted within an acid resistant, cylindrical housing which is inserted permanently within the animals stomach [38].

In combination with herd management software, RFID animal identification systems can include detailed information like records of medical treatments, animal growth performance data, pasture performance data, movement of animals, purchase and sale dates, and carcass feedback data. Also, through the use of electronic animal identification subsidies based on the number of animals or their genetic background can be allocated properly, electronic feeding stations can be implemented or tracing back of stolen stock [39].

Animal Identification Programs

Australia was the pioneer in the implementation of mandatory identification systems, developing its own individual whole-of-life traceability program for livestock; the so called National Livestock Identification Scheme (NLIS). This goal of the program is to be a "... *permanent whole-of-life identification system that enables individual animals to be tracked from property of birth to slaughter for food safety, product integrity and market access purposes*". NLIS requires all calves to have RFID devices before leaving the property on which they were born. These RFID devices can be either ear tags or rumen bolus/ear tag combinations [40].

In USA, the USDA (United States Department of Agriculture) established in 2005, the National Animal Identification System (NAIS) as a voluntary program that covered Camelids (llamas and alpacas), cattle and bison, Cervids (deer and elk), Equines, Goats, Poultry, Sheep and Swine. The goal of NAIS was "*to be able to identify all animals and premises that have had contact with a foreign or domestic animal disease of concern within 48 hours after discovery*". Recently, the 5th February, 2010, USDA announced a new framework for animal disease traceability in the United States. The framework will apply only to animals moved in interstate commerce, will be administered by the States and Tribal Nations to provide more flexibility and will not be mandatory. The program will include RFID as one of the official devices for animal identification [41].

In 2003, the Canadian Cattle industry committed to the transition to RFID technology to ensure Canada's Cattle Identification Program continues its role as a trace back system. The Canadian Cattle Identification Program (CCIP) is an industry-led trace-back system designed to help trace sources of animal health and food safety problems. It was introduced in 2001, and is applicable to all cattle and bison in Canada. All tags are embedded with a unique identification number that is allocated by the Canadian Cattle Identification Agency (CCIA), which collects and stores identification information for most provinces in its national database [42].

In Europe, the legislation takes into account the electronic identification of livestock as a tool to improve traceability of livestock and products thereof. From the first of January 2010, sheep and goats already need to be electronically identified [43]. Bovine animals are currently identified by two plastic eartags. The European Commission is currently exploring the possibility of introducing electronic identification as official method to identify bovine animals within the EU [44].

2.3. Other Livestock Applications

Temperature is one the most important parameter to monitor in livestock, because temperature fluctuations to be a great indicator of health problems in livestock. If any unusual temperature readings arise, then a farmer can be notified and take appropriate actions, such as removing this animal from the rest and checking it for illness. RFID has been used as a new technique for measuring core body temperature that are minimally invasive and provide continuous, remote, real-time information. RFID sensing devices can be injected into the animal (under the skin), and provide temperature readings when interrogated by an RFID reader. Application tests have been performed in horses [45], poultry, beef and dairy cattle, showing good accuracy, resolution and response time for temperature measurement [46]. Manufacturers are looking to improve this technology in the future, in order to provide information on an animal's hormonal changes, blood pressure and even possibly disease identification [47]. Also, the chewing and ruminating behaviors can be studied by the implementation of wireless automatic systems, addressing the dietary factors affecting normal rumen function of dairy cows [48].

2.4. Precision Agriculture

The development of RFID applications in precision agriculture makes possible to increase efficiencies, productivity and profitability while minimizing unintended impacts on wildlife and the environment, in many agricultural production systems. The chain, from harvest to the consumer's plate, forms a single entity. Thus, the documentation of production using advanced electronic equipment can provide valuable information for knowing the complete history of the products [49]. Moreover, the real time information from the fields will provide a solid base for farmers to adjust strategies at any time. Instead of take decisions based in some hypothetical average condition, which may not exist anywhere in the reality, a precision farming approach recognizes differences and adjusts management actions accordingly [50].

Climate Monitoring

Wireless sensing has become an important issue in environmental monitoring. The relatively low cost of the devices allow the installation of a dense population of nodes that can adequately represent the variability present in the environment. They can provide risk assessment information, like for example alerting farmers at the onset of frost damage and providing better microclimate awareness. The automation of the monitoring process can be used in diverse types of climates and conditions. Hamrita and Hoffacker (2005) developed a lab prototype for wireless measurement of soil temperature. The system was based in a commercial 13.56 MHz RFID tag. Measurements showed a high correlation (greater than 99%) with those obtained using a thermocouple [51].

Precision Irrigation

Efficient water management is a major concern in many cropping systems. RFID have a big potential for represent the inherent soil variability present in fields with more accuracy than the current systems available. Thus, the benefit for the producers is a better decision support system that allows to maximize their productivity while saving water. Also, RFID eliminates difficulties to wire sensor stations across the field and reduces maintenance cost. Since installation of RFID is easier than existing wired solutions, sensors can be more densely deployed to provide local detailed data. Instead than irrigating an entire field in response to broad sensor data, each section could be activated based on local sensors. Following this approach, Vellidis et al. (2008) developed a prototype of smart sensor array for scheduling irrigation in cotton. The array consists of a centrally located receiver connected to a laptop computer and multiple sensor nodes installed in the field. The nodes consist of sensors (up to three soil moisture sensors and up to four thermocouples), a specially designed circuit board, and a RFID tag which transmits data to the receiver [52].

Greenhouses

Yang et al. (2008) reported a multi-functional remote sensing system that integrates RFID technology with spectral imaging and environmental sensing in a greenhouse. The multi-spectral imaging system was used for remote sensing of the canopy of cabbage seedlings. Greenhouse temperature, relative humidity, and lighting conditions were measured above the crop [53].

Farm Machinery

RFID implemented in off-road vehicles, such as tractors or combine harvester, could allow exchanging data with static infrastructure or with other vehicles, creating mobile RFID systems. Attaching RFID to the products (seeds, fertilizers, pesticides, etc.) and installing readers in the machinery (tractors, implements, self-propelled machines), would enable to record what is being put into the implement's hopper or tank. Transparency is gained for the purpose of quality assurance, knowing which fertilizer was spread or when, and which pesticides or insecticides were used. In the case of a chemical sprayer, additional information such as requirements for PPE (Personal Protective Equipment) and COSHH (Control of Substances Hazardous to Health) regulations could also be added to the tag. The information saved in the computer, in a standard format would be transferred and used in combination of decision support systems [54].

2.5. Cold Chain Applications

Every day, millions of tons of temperature sensitive goods are produced, transported, stored or distributed worldwide. For all these products the control of temperature is essential. The term "cold chain" describes the series of interdependent equipment and processes employed to ensure the temperature preservation of perishables and other temperature-controlled products from the production to the consumption end in a safe, wholesome, and good quality state [55]. The supply chain management of fresh foods requires fast decisions because goods are forwarded within hours after arrival at the distribution center. The quality of fresh meat, fish, or agricultural products might change rapidly. Appropriate planning calls for more information than that which could be provided by standard RFID tracking and tracing.

Several applications for monitoring cold chain logistics by means of RFID have been reported. The majority are oriented to perishable food products. Here are the most representative to our knowledge.

The use of microbial growth models combine with information from active RFID has been faced. These models allow the prediction of microbiological safety and quality of foods, by monitoring the environment without recourse to further microbiological analysis. Thus immediate decisions on the quality and/or safety of fresh produce can be made based on the temperature profile of the supply chain. Three different cases were studied: frozen dairy product, meat carcass chilling and fermented meat processing [56].

An important step in cold chain management is recording the temperature throughout the supply chain. Implementation of HACCP (Hazard Analysis and Critical Control Points) requires measurements to ensure that the prescribed control limits are not being violated. Ogasawara and Yamasaki (2006) reported a cold chain solution that uses RFID tags with embedded temperature sensors. It also introduced a temperature-managed traceability starter kit that contributes to effective risk management by easily enabling consistent temperature management throughout transportation processes [57].

The integrity of the cold chain must be maintained from the very beginning of production or processing, through each link (loading, unloading, transport, handling, storage) to the consumer end. Gras (2006) monitored a cold chain of frozen goods using semi-passive and active RFID instrumented with temperature sensors. The experimental work covers four steps of the cold chain: production, transportation, storage and delivery. Data was linked with computerized cold chain management system [58].

Environmental temperature can differs from each other depending in the location of the logger, packing material or heat dissipation of the product [59, 60]. RFID tags can be also used to measure, not just the walls of the vehicle, but also inside the boxes. Amador et al. (2008) showed the use of RFID for temperature tracking in an international shipment of pineapples from a packing house, in Costa Rica, to a wholesale storage, in USA. They studied the use of RFID in temperature monitoring by comparing the performance of RFID temperature tags versus conventional temperature tracking methods, as well as RFID temperature tags with probe versus RFID temperature tags without probes and their utilization along the supply chain. The temperature mapping of a shipping trial comprising pallets of crownless pineapples instrumented using different RFID temperature dataloggers and traditional temperature dataloggers and packed in two kinds of packages (corrugated boxes and reusable plastic containers) inside a container was performed. The results showed that RFID temperature tags are analogous with regards to accuracy to the conventional methods, but have a superior performance because they allow quick instrumentation and data recovery and the possibility of accessing the sensor program and data at any point of the supply chain without line of sight [61].

The fresh fish logistic chain has been also monitored using RFID. Abad et al. (2009) validated a RFID smart tag instrumented with light, temperature and humidity sensors. The system provides real-time traceability information of the product to the different fish distribution chain links. The RFID tag was placed on a corner of the fish box to ensure a correct real-time reading of the temperature and relative humidity measurements (maximum reading distance of about 10 cm) from the outside without the need of opening the fish box [11].

The accuracy of data loggers is a critical issue in cold chain management. This accuracy becomes even more important if the objective is early detection of temperature changes and gradients. Standards for food distribution allow deviations of $\pm 0.5^{\circ}C$ from the set point [62]. Jedermann *et al.* did a comparison of three different types RFID loggers in a climatic chamber, finding that the percentage of measurements with a difference to the average value less than the deviation $\pm \delta$ was between 66% and 73% [12].

Demands for mixed loads of products require different storage temperatures and the trend of refrigerated transport is to use multi-compartmental vehicles. Jedermann et al. (2009) monitored 16 multi-compartmental trucks using semi-passive RFID instrumented with temperature sensors (Turbo Tag) detecting temperature gradients. The authors concluded that semi-passive tags can be used to monitor environmental variables such as the temperature of chilled food refrigerated goods, to identify problem areas and to raise alarms. RFID loggers are good tools are cost-effective for the characterization of refrigerated transport units such as trucks or containers [12].

Intermodal Transport of Agro-food Products

In the intermodal transportation, the performance of radio waves inside metal enclosed areas was studied. Furh and Lau (2005) tested a radio frequency device in a metal cargo container and demonstrated that it is possible to communicate with the outside world [63]. Jedermann et al. (2006) presented a system for intelligent containers combining wireless sensor networks and RFID [64]. Laniel et al. (2008) focuses on the 3-D mapping of RFID signal strength inside a 12m (40') refrigerated marine container. Three different types of radio frequency configurations were tested: 2.4GHz, 915MHz and 433MHz.Tests were performed with an empty container and the main goal was to find a frequency and configuration that would allow real time reading of the temperature in a shipment of perishable products using RFID. Results obtained in this study showed that wave propagation inside a closed marine container is significantly higher at 433MHz than at 915MHz or 2.4GHz, with attenuation averages of 19.57 and 18.20 versus 36.49 and 35.91 and 29.91 and 29.78 dBm respectively [65]. At 433 MHz the wavelength is approximately a meter, enabling signals to diffract around obstructions. The level of diffraction depends on the size of the object versus the signal wavelength. At 2.4GHz the diffraction is very limited and therefore not recommended for most cold chain applications which are in crowded environments [66].

3. CHALLENGES AND LIMITATIONS

RFID itself is not a very new technology, but its commercial use is very recent. Thus, implementing RFID involves a multitude of challenges. A significant proportion of RFID deployments remain exploratory. There is a need to know the long-term behavior of the systems. Most of the applications reported have short experimental periods. Longer testing and experimentation is necessary for validate some of applications presented.

In Agriculture applications, RFID is often exposed to harsh environments with excessive dirt, dust, moisture, and temperature extremes. They must function in both extreme heat and cold, from -30 to 70 degrees. Haapala (2003) validated the performance of electronic identification tags for animals, under extremely cold temperature (-25 ºC) [67]. Moreover, in the food industry tags should withstand the pasteurization process, boiling point temperatures, x-ray and gamma radiation, which is commonly used for sterilization. Advances in RFID make the technology more useful to food processors with new developments, like the gamma sterilizable tags now available, with a resistant up to 500 kilogray (kGy) of gamma energy [68].

One limitation might be that these monitoring systems create huge volumes of data that are difficult to manage, causing a huge increase in the daily volume of data in a corporate information technology system. An RFID implementation can generate 10 to 100 times more information than traditional barcode technology. Database administrators need to be able to deal with the potential stresses on the databases, both in terms of speed and volume. Even so, data volume can be overwhelming to the network. If a product have 1,000 bytes of data associated with it, the RFID monitoring system would generate 10 terabytes of data per year. If the data of five years is stored, that means a database of 50 terabyte. Thus, the solution lies in implementing a decentralized data management system. Data can be pre-processed and duplicate information eliminated close to their point of origin by intelligent systems, which could be sited at the level of the tag or reader [69, 70].

The read range performance of tags differs extraordinarily. Radio propagation in real environments is complex due to multipath propagation, shadowing and attenuation. In agriculture, the radio frequency faces challenges due to placement of nodes for wide-area mesh coverage and reliable link quality above crop canopies. RFID must be able to operate in a wide range of environments such as bare fields, vineyards, orchards, from flat to complex topography and over a range of weather conditions, all of which affect radio performance [71]. In these situations, the link power budget is dependent on crop growth and terrain in addition to more common factors such as node spacing and antenna height [72]. For applications inside buildings like burns, greenhouses or warehouses, the radio signal has to go through many objects like walls, windows, pallets, machines, etc. which also cause a significant reduction in signal strength. In general, the higher the frequency, the longer the communication range, and the faster the communication, this means that more data can be transmitted. In near field applications where is possible to have lots of tags it might be better to have a tag with low read range to avoid false positive reads. In other applications, like crop monitoring, it is better to choose a long read range to optimize the read performance.

An important research topic that must be faced is fault detection and isolation. In a remote sensing application is essential to detect the erroneous measurements. False reads can be done as a result of radio waves being distorted, deflected, absorbed, and interfered with. Wrong information provided by the monitoring system should be identified and skipped. Also the implementation of artificial intelligence in the core of the system can block the transmission of erroneous data [25].

RFID data loggers are available in high quantities, but they require manual handling because of their low reading range. Another disadvantage is that temperature loggers are only available for the 13.56 MHz HF-Range. The major drawback of this band is the limited reading range of about 20 cm. If a gate reader scans items automatically upon arrival at the warehouse, the reading range has to cover several meters. Also these tags take around five seconds to transfer recorded temperature values over the RFID interface [12]. A higher data rate is required according to a normal flow of goods in a warehouse.

Another important issue is to deal with the physical limitations of RFID. Metals and liquids inhibit the propagation of electromagnetic waves. This is particularly true for UHF and microwave frequencies (2.4GHz). Some temperature sensitive products such as fruits, vegetables or juices have high water content, sometimes more than 90%. As a result, performance can be affected by the item on which the tag is attached [25, 69]. However, reflections and product dimensions is important for liquid and dielectric materials as well since much of the power loss occurs at the interfaces between air and the medium [73]. Moreover, the last generation of Class 4 RFID tags can be configured in a mesh network. In this type of network, the tags can communicate each other to get to a reader circumventing environmental obstacles and extend the size of the system [74].

The lack of uniformity in global standards makes the RFID implementations more difficult. Managing multiple readers and related hardware can be a challenge, especially across multiple facilities. There is practically no part of the spectrum available worldwide because governments have assigned different uses for the various parts of this spectrum, with the exception of the ISM (industrial, scientific and medical) bands. In order to solve this problem, the Auto-ID Center has explored the concept of "agile" readers that will allow the network to operate at different frequencies in a wide variety of geographical locations [75].

The level of granularity is a limitation in most of the applications. Normally three levels of granularity are considered: pallet, case or item-level. The primary advantage of case or item-level tagging over pallet-level tagging is more detailed and accurate information, since each pallet in a load and each carton on a pallet can experience temperature variations. Instead of reject an entire shipment goods can be considered on a pallet-by-pallet or case-by-case basis. But high granularity also means much more tags to handle with, higher costs and huge data to be processed [25].

The lack of skilled personnel is another limitation in many agricultural implementations. Many companies have no qualified personnel for this purpose; there is a shortfall between the supply of talent and the market demand. The expansion of RFID in agriculture requires more agriculture engineers, computer scientists and technicians with RFID skills.

Information sharing is probably one of the greatest challenges, but is essential for achieving a trustable and efficient traceability control in Agriculture. Obtaining the required level of trust and cooperation across the supply chain, collaboration with supply chain partners both up and down the chain, is necessary. However, there is a strong resistance to share information on applications that depend on data from various trading partners, information sharing issues must be resolved to achieve maximum benefit.

One of the current challenges in smart tags is the integration of chemical sensors onboard of flexible tags [11] to monitor for example the ripening or deterioration gases generated by food products. In the case of fruit logistics, volatile compounds like ethanol and ethylene are very important to detect and quantify [76]. Resistive sensors such as Metal Oxide Sensors (MOS) for volatile evaluation have been developed into commercial MEMS by means of the development of Ultra Low Consumption Hot plates which allow the reduction of the size of the sensor and thus the power required for proper operation. Micro Hot Plate temperature is controlled from ambient to 500°C with a thermal efficiency of 8ºC/mW and thermal response time of 0.6 ms. The fabrication methodology allows integration of an array of gas sensors of various films with separate temperature control for each element in the array, and circuits for a low-cost MOS-based gas sensor system. But this technology has it main drawback in the lack of specificity of sensor [77].

Vergara et al. (2006) developed an RFID reader with onboard micro-machined metal oxide gas sensors aimed at monitoring concentration of gases, such as acetaldehyde or ethylene during fruit transport and vending. The developed platform integrates a commercial off the shelf inductive coupling RF transceiver in the 13.56MHz band, fully compliant with the ISO15693 standard, micro-hotplate gas sensors, driving and readout electronics [18].

4. Conclusion

As it was shown, the applications of RFID in agriculture are many and varied. The use of RFID in agriculture and food industry provides new features that have the potential to be an economically viable, speeding up operations and improving data accuracy. The value of technology can be best realized when integrated with agronomic knowledge, using the information gathered in the improvement of decision support systems. Also improving operations by providing early warning of equipment failure and a predictive maintenance tool, improving energy management, providing automatic record-keeping for regulatory compliance, eliminating personnel training costs or reducing insurance costs. The collaboration and synergy of sensing, processing, communication and actuation is the next step to exploit the potential of these technologies.

From 2004 to 2010 the evolution of RFID technology has been developed very fast, adding new features to traditional automatic identification and data capture applications. Semi-passive tags can be used to monitor environmental variables, such as the temperature, to identify problem areas and to raise alarms. RFID loggers are good tools that are available in high quantities and are cost-effective. However, they require manual handling because of their low reading range.

An important benefit of the systems is the visibility that it can give along the food chain. Measurements obtained are consistent and provide valuable information on the conditions encountered during the life cycle of the products. It is possible to address, at regular time increments, what is happening with the product. Another advantage is providing effective support in legal situations as well as safety inspections.

However, a significant proportion of RFID deployments remain exploratory and there are important challenges to face such as data management, read range performance, fault detection and isolation or the level of granularity.

References

[1] ISO, 11784 Radio frequency identification of animals – Code structure. In International Organization for Standardization: 2004.

[2] ISO, 11785 Radio frequency identification of animals – Technical concept In International Organization for Standardization: 1996.

[3] ISO, 14223 Radiofrequency identification of animals – Advanced transponders. In International Organization for Standardization: 2003.

[4] ISO/IEC, 7816 Identification cards – Integrated circuit cards. In International Organization for Standardization / International Electrotechnical Commission: 2008.

[5] ISO/IEC, 14443 Identification cards – Contactless integrated circuit cards – Proximity cards. In International Organization for Standardization / International Electrotechnical Commission: 2008.

[6] ISO/IEC, 15693 Identification cards – Contactless integrated circuit cards – Vicinity cards In International Organization for Standardization / International Electrotechnical Commission: 2008.

[7] ISO/IEC, 18000 Information technology – Radio frequency identification for item management. 2008.

[8] Trevarthen, A., The National Livestock Identification System: The Importance of Traceability in E-Business. *Journal of Theoretical and Applied Electronic Commerce Research* 2007, *2*, 49-62.

[9] Barbari, M.; Conti, L.; Simonini, S., Spatial Identification of Animals in Different Breeding Systems to Monitor Behavior. In *Livestock Environment VIII*, ASABE, Ed. Iguassu Falls, Brazil, 2008.

[10] Akyildiz, I.F.; Su, W.; Sankarasubramaniam, Y.; Cayirci, E., Wireless sensor networks: a survey. *Comput. Netw.* 2002, *38*, PII S1389-1286(01)00302-4.

[11] Abad, E.; Palacio, F.; Nuin, M.; González de Zárate, A.; Juarros, A.; Gómez, J.M.; Marco, S., RFID smart tag for traceability and cold chain monitoring of foods: Demonstration in an intercontinental fresh fish logistic chain. *J. Food. Eng.* 2009, *93*.

[12] Jedermann, R.; Ruiz-Garcia, L.; Lang, W., Spatial temperature profiling by semi-passive RFID loggers for perishable food transportation. *Comput. Electron. Agric.* 2009, *65*, 145-154.

[13] Amador, C.; Emond, J.P.; Nunes, M.C., Application of RFID Technologies in the Temperature Mapping of the Pineapple Supply

Chain. In *Food Processing Automation Conference*, Providence, Rhode Island (USA), 2008.

[14] Chang, K.; Kim, Y.H.; Kim, Y.; Yoon, Y.J., Functional antenna integrated with relative humidity sensor using synthesised polyimide for passive RFID sensing. *Electron. Lett.* 2007, *43*, 259-260.

[15] Todd, B.; Phillips, M.; Schultz, S.M.; Hawkins, A.R.; Jensen, B.D., Low-Cost RFID Threshold Shock Sensors. *IEEE Sens. J.* 2009, *9*, 464-469.

[16] Cho, N.; Song, S.J.; Kim, S.; Yoo, H.J., A 5.1-mu W UHF RFID tag chip integrated with sensors for wireless environmental monitoring. *Esscirc 2005: Proceedings of the 31st European Solid-State Circuits Conference* 2005, 279-282.

[17] Murković, I.; Steinberg, M.D., Radio Frequency tag with optoelectronic interface for distributed wireless chemical and biological sensor applications. *Sensor. Actuat. B-Chemical.* 2009, *In press*.

[18] Vergara, A.; Llobet, E.; Ramírez, J.L.; Ivanov, P.; Fonseca, L.; Zampolli, S.; Scorzoni, A.; Becker, T.; Marco, S.; Wöllenstein, J., An RFID reader with onboard sensing capability for monitoring fruit quality. In *Eurosensors 2006*, Goteborg, Sweden, 2006.

[19] Wentworth, S.M., Microbial sensor tags. In *2003 IFT (The Institute of Food Engineering) Annual Meeting*, Chicago, Illinois, USA, 2003.

[20] Beulens, A.J.M.; Broens, D.F.; Folstar, P.; Hofstede, G.J., Food safety and transparency in food chains and networks - Relationships and challenges. *Food Control* 2005, *16*, 481-486.

[21] Clapp, S., *A brief history of traceability*. 2002.

[22] Bechini, A.; Cimino, M.; Marcelloni, F.; Tomasi, A., Patterns and technologies for enabling supply chain traceability through collaborative e-business. *Information and Software Technology* 2008, *50*, 342-359.

[23] Regattieri, A.; Gamberi, M.; Manzini, R., Traceability of food products: General framework and experimental evidence. *J. Food. Eng.* 2007, *81*, 347-356.

[24] Jones, P., Clarke-Hill, C., Shears, P., Comfort, D., Hillier, D., Radio frequency identification in the UK: opportunities and challenges. *International Journal of Retail & Distribution Management* 2004, *32*, 164-171.

[25] Angeles, R., RFID technologies: Supply-chain applications and implementation issues. *Inform. Syst. Manage.* 2005, *22*, 51-65.

[26] Twist, D.C., The Impact of Radio Frequency Identification on Supply Chain Facilities. *J Facilities Manage.* 2005, *3*, 226-239.

[27] Attaran, M., RFID: an enabler of supply chain operations. *Supply. Chain. Manag.* 2007, *12*, 249-257.

[28] Ngai, E.W.T.; Cheng, T.C.E.; Au, S.; Lai, K.H., Mobile commerce integrated with RFID technology in a container depot. *Decision Support Systems* 2007, *43*, 62-76.

[29] Meyer, G.G.; Främling, K.; Holmström, J., Intelligent Products: A survey. *Comput. Ind.* 2009, *60*, 137-148.

[30] Ruiz-Garcia, L.; Steinberger, G.; Rothmund, M., A Model and Prototype Implementation for Tracking and Tracing Agricultural Batch Products along the Food Chain. *Food Control* 2009, *21*, 112-121.

[31] Scheer, F.P., Optimising supply chains using traceability systems. In *Improving traceability in food processing and distribution*, limited, W. p., Ed. Cambridge, England, 2006; pp 52-64.

[32] Koutsoumanis, K.; Taoukis, P.S.; Nychas, G.J.E., Development of a safety monitoring and assurance system for chilled food products. *International Journal of Food microbiology* 2005, *100*, 253-260.

[33] Emond, J.P.; Nicometo, M., Shelf-life prediction and FEFO inventory management with RFID. In *Cool Chain Association Workshop. Temperature measurements-when, where and how?*, Knivsta, Sweden, 2006.

[34] Yam, K.L.; Takhistov, P.T.; Miltz, J., Intelligent packaging: Concepts and applications. *Journal of Food Science* 2005, *70*, R1-R10.

[35] Roberti, M. Beware of RFID's Hysteresis Effect.

[36] Swedberg, C. Cattle Auctioneer Promotes Tracking Plan. http://www.rfidjournal.com/article/view/1655 (10-January-2010).

[37] Munak, A., Preface. In *CIGR Handbook of Agricultural Engineering Volume VI Information Technology*, CIGR, Ed. 2006; pp XVII-XVIII.

[38] Finkenzeller, K., *RFID Handbook radio-frequency identification fundamentals and applications*. 2nd edition ed.; England, 2004.

[39] Voulodimos, A.S.; Patrikakis, C.Z.; Sideridis, A.B.; Ntafis, V.A.; Xylouri, E.M., A complete farm management system based on animal identification using RFID technology. *Comput. Electron. Agric.* 2010, *70*, 380-388.

[40] Tonsor, G.T.; Schroeder, T.C. *Australia's Livestock Identification Systems: Implications for United States Programs*; 2004.

[41] APHIS-USDA. Animal Disease Traceability. http://www.aphis.usda.gov/traceability/

[42] CCIA Canadian Cattle Identification Agency. http://www.canadaid.com (22-January-2010).

[43] European Commission, Commission Regulation (EC) No 759/2009 of 19 August 2009 amending the Annex to Council Regulation (EC) No 21/2004 establishing a system for the identification and registration of ovine and caprine animals. In 2009.

[44] Arcadia International *Study on the introduction of electronic identification (EID) as official method to identify bovine animals within the European Union*; European Commission: 2009.

[45] Marsh, J.R.; Gates, R.S.; Day, G.B.; Aiken, G.E.; Wilkerson, E.G., Assessment of an Injectable RFID Temperature Sensor for Indication of Horse Well-Being. In *ASABE Annual International Meeting 2008*, Rhode Island, USA, 2008.

[46] Brown-Brandl, T.M.; Yanagi, T.; Xin, H.; Gates, R.S.; Bucklin, R.A.; Ross, G.S., A new telemetry system for measuring core body temperature in livestock and poultry. *Appl. Eng. Agric.* 2003, *19*, 583-589.

[47] Hostettor, J. Animal-tracking chips now let you in on how Fido is feeling. http://www.usatoday.com/tech/news/techinnovations/2003-04-21-animal-chip_x.htm (2-Feb-2010).

[48] Kononoff, P.J.; Lehman, H.A.; Heinrichs, A.J., Technical note - A comparison of methods used to measure eating and ruminating activity in confined dairy cattle. *Journal of Dairy Science* 2002, *85*, 1801-1803.

[49] Auernhammer, H., The Role of Mechatronics in Crop Product Traceability. Agricultural Engineering International: the CIGR Journal of Scientific Research and Development. In *Club of Bologna meeting*, Chicago, IL., USA, 2002.

[50] USC Precision Agriculture. http://www.gpoaccess.gov/uscode/index.html (14/02/2009).

[51] Hamrita, T.K.; Hoffacker, E.C., Development of a "smart" wireless soil monitoring sensor prototype using RFID technology. *Appl. Eng. Agric.* 2005, *21*, 139-143.

[52] Vellidis, G.; Tucker, M.; Perry, C.; Wen, C.; Bednarz, C., A real-time wireless smart sensor array for scheduling irrigation. *Comput. Electron. Agric.* 2008, *61*, 44-50.

[53] Yang, I.-C.; Chen, S.; Huang, Y.-I.; Hsieh, K.-W.; Chen, C.-T.; Lu, H.-C.; Chang, C.-L.; Lin, H.-M.; Chen, Y.-L.; Chen, C.-C.; Lo, Y.M., RFID-integrated Multi-Functional Remote Sensing System for Seedling

Production Management. In *2008 ASABE Annual International Meeting*, Providence, Rhode Island, USA, 2008.

[54] Watts, A.; Miller, P., Applications of RFID (radio frequency identification) in agriculture. *Pesticide Outlook* 2002, *13*, 254 - 258.

[55] Zhang, L. Cold Chain Management. Cranfield University, 2007.

[56] McMeekin, T.; Smale, N.; Jenson, I.; Tanner, D. In *Microbial growth models and temperature monitoring technologies*, 2nd international Workshop Cold Chain Management, Bonn, Germany, 8-9 May 2006, 2006; Kreyenschmidt, J. P., B., Ed. Bonn, Germany, 2006; pp 71-78.

[57] Ogasawara, A.; Yamasaki, K., A temperature-managed traceability system using RFID tags with embedded temperature sensors. *NEC Technical Journal* 2006, *1*, 82-86.

[58] Gras, D. In *RFID based monitoring of the cold chain*, 2nd international Workshop Cold Chain Management,, Bonn, Germany, 2006; Kreyenschmidt, J. P., B., Ed. Bonn, Germany, 2006; pp 81-82.

[59] Moureh, J.; Laguerre, O.; Flick, D.; Commere, B., Analysis of use of insulating pallet covers for shipping heat-sensitive foodstuffs in ambient conditions. *Comput. Electron. Agric.* 2002, *34*, PII S0168-1699 (01)00181-8.

[60] Raab, V.; Bruckner, S.; Beierle, E.; Kampmann, Y.; Petersen, B.; Kreyenschmidt, J., Generic model of shelf life dynamics in support of cold chain management in pork and poultry supply chains. *J. Chain. Network. Sci.* 2008, *8*, 59-73.

[61] Amador, C.; Emond, J.P.; Nunes, M.C.N., Application of RFID technologies in the temperature mapping of the pineapple supply chain. *Sensing and Instrumentation for Food Quality and Safety* 2009, *2009*, 26-33.

[62] CEN, 12830. Temperature recorders for the transport, storage and distribution of chilled, frozen, deep-frozen/quick-frozen food and ice cream - Tests, performance, suitability. In European Committee for Standardization: 1999; Vol. EN 12830.

[63] Fuhr, P.; Lau, R., Mesh radio network performance in cargo containers. *Sensors Magazine Online* 2005.

[64] Jedermann, R.; Behrens, C.; Westphal, D.; Lang, W., Applying autonomous sensor systems in logistics - Combining sensor networks, RFIDs and software agents. *Sensor. Actuat. A-Physical.* 2006, *132*, 370-375.

[65] Laniel, M.; Emond, J.P.; Altunbas, A.E., RFID Behavior Study in Enclosed Trailer/Container for Real Time Temperature Tracking. In

Food Processing Automation Conference, Providence, Rhode Island (USA), 2008.

[66] Technologies, S. Selecting the Right Active Frequency. http://www.autoid.org/2002_Documents/sc31_wg4/docs_501-520/520_18000-7_WhitePaper.pdf (5-September-2009).

[67] Haapala, H.E.S., Operation of RFID in Cold Environment. In *Livestock Environment VIII*, ASABE, Ed. Iguassu Falls, Brazil, 2008.

[68] Andrechak, G.; Wiens, R.A. Hitachi μ-chip RFID Technology Compatible with Gamma Sterilization. http://www.nordion.com/documents/uChip-Gamma-White-Paper.pdf (02-Feb-2010).

[69] Ruiz-Garcia, L. Development of Monitoring Applications for Refrigerated Perishable Goods Transportation. Universidad Politécnica de Madrid, Madrid, 2008.

[70] Roberti, M., Seven Reasons to Act Now. *RFID Journal* 2003.

[71] Andrade-Sanchez, P.; Pierce, F.J.; Elliot, T.V., Performance Assessment of Wireless Sensor Networks in Agricultural Settings. In *2007 ASABE Annual International Meeting* Minneapolis, Minnesota (USA), 2007.

[72] Tate, R.F.; Hebel, M.A.; Watson, D.G., WSN Link Budget Analysis for Precision Agriculture. In *2008 ASABE Annual International Meeting*, ASABE, Ed. Providence, Rhode Island, RI, USA, 2008.

[73] Fletcher, R.; Marti, U.P.; Redemske, R., Study of UHF RFID signal propagation through complex media. In *Antennas and Propagation Society International Symposium, 2005 IEEE*, 2005; Vol. 1B, pp 747-750.

[74] Engels, D.W.; Sarma, S.E. Standardization Requirements within the RFID Class Structure Framework. http://www.autoidlabs.org/uploads/media/AUTOIDLABS-WP-SWNET-011.pdf (05-01-2010).

[75] Haller, S.; Hodges, S. The need for a universal smart sensor network. www.autoidcenter.org (7-September-2009).

[76] Barreiro, P., Robla, J. I., Rodríguez-Bermejo, J., Ruiz-Garcia, L., Ruiz-Altisent M., Desarrollo de sistemas para la supervisión multidistribuida de la carga en transportes frigoríficos intermodales. *Fruticultura profesional* March 2005, 2005, pp 44-51.

[77] Wise, K.D., Integrated sensors, MEMS, and microsystems: Reflections on a fantastic voyage. *Sensor. Actuat. A-Physical.* 2007, *136*, 39-50.

In: Radio Frequency Identification ISBN 978-1-61122-416-0
Editor: Alison R. McAdams, pp. 55-74

Chapter 3

PLANTS WITH IMPLANTED RFID MICROCHIPS: TRACEABILITY AND OUTLOOK IN INFORMATION MANAGEMENT SYSTEMS

A. Luvisi[1,*] and M. Pagano[2,†]

[1]Department of Tree Science, Entomology, and Plant Pathology "G. Scaramuzzi", University of Pisa, Via del Borghetto, 80, 56124 Pisa, Italy

[2]Department of Crop, Soil and Environmental Science, University of Florence, Viale delle Idee, 30, 50019 Sesto Fiorentino (FI), Italy

ABSTRACT

In recent years consumers have demanded safer and more wholesome products, and stricter regulations have supported these expectations. Thus, the need to know more about the origins and qualitative characteristics of food products commercialized worldwide has increased. Traceability is essentially the "history" of a product, from its origin to the shelf, and in agriculture traceability reaches back to the genetic status of products, whether they are of animal or vegetal origin. In

---

* E-mail: aluvisi@agr.unipi.it
† E-mail: mario.pagano@unifi.it

order to verify the status of a product, different tools are used during different phases of the production line.

In agriculture, radio-frequency identification (RFID) technology has been introduced efficiently in animal identification systems for traceability purposes, and legal regulations have been put in place for this sector. Differently from animals, where the microchip is often inserted within the organism, RFID applications for plants mainly regard food traceability, logistics or harvesting, in which the microchip (tag) is not inserted inside the plant or product.

In the last 5 years, experimental trials focusing on inserting tags within plants have been carried out on *Citrus* spp., *Cypress* spp., *Platanus* spp., *Vitis* spp. and other genus. Keeping in mind plant histology and organ size, different techniques and tag allocations have been proposed, yet the standardization of RFID tagging in plants does not seem possible. In fact, specific solutions have been suggested for tagging plants with respect to growth stage, anatomy or aesthetic considerations. Moreover, even the aim of tagging can orientate the technology used and, consequentially, the methods of tag insertion. Nowadays, identification of mother plants used for propagative purposes or plant pathology monitoring represent the most relevant cases of study or practical applications, but there are also some interesting outlooks with regard to RFID integration with precision farming and for developing information management systems.

Implanting an RFID microchip inside a plant represents the first step for developing integrated systems that involve positioning techniques, mobile or Wi-Fi devices and a Web 2.0 approach. These complex systems can strongly support farm management, differentiate the final product in markets, and help controllers in their verification and increase consumer confidence in product origin.

1. Introduction

1.1. Certification and Traceability in Agriculture

Nowadays plant identity—understood as the entire genetic, phenotypic and health characteristics of the plant—is not only a matter of discussion for plant growers or researchers but it is also of interest, more or less directly, for the whole society. The need for knowledge about the origins and qualitative characteristics of food products or plants commercialized worldwide has increased due to consumer demands which are oriented toward safe and wholesome products [1], also due to recent negative events regarding the food sector which have in turn led to stricter regulations for better public health. Beginning in the 1960s, in an essential step to guarantee quality for users, the former European Economic Community was involved in defining legal regulations regarding the health status of grapes (68/93/EEC), ornamental plants (77/93/EEC), and fruit trees (92/34/EEC); a similar approach was undertaken by the North American Plant Protection Organization (NAPPO) [2]. Regulatory dispositions followed the general trends in agriculture in the past century: not only new farming approaches and consequential environmental impact, but also worldwide trade of products posing new challenges to import/export regulations. Food safety, market protection, and ecological conservation are common themes of discussion that are closely linked to the products we find on our plates and in our glasses: to know what it is, who produced it, where it comes from and where it was processed are no longer questions asked only by food lovers, but also by mass-market consumers and government agencies. With regard to plants or products deriving from them, this approach is even more relevant when foodstuffs are involved: product safety is the fundamental factor making traceability important as recent studies on food safety have shown that approximately seven million people a year are affected by food-borne illness [3]. These concepts were reinforced by the EU through ''The European White Paper on Food Safety" [4] and by the FAO in ''The Bangkok Declaration and Strategy on Aquaculture Development" [5].

Many foods or agricultural products have to carry identifying labels or documents, as reported by legal regulations (i.e., 2000/13/EC; Title 21 of CFR, Food and Drug Administration). The proposed regulations seem to efficiently respond to the main concerns, especially with regard to health or ecological issues, and a more-in-depth system for identification and traceability can appear pleonastic. However in some cases (i.e., Parmigiano Reggiano [6] or grapevine [7]), a more specific and detailed traceability system that addresses each single factor—such as the cheese or grapevine plant itself—can provide useful information, particularly with regard to the quality of final product. In fact, labels or documents are not always associated with products such as plants, or if they are physically attached to them, their loss or removal — voluntary or not—can occur, making identification difficult or impossible or interrupting traceability. A product traceability system is fundamentally based on product identification, data to trace, product routing, and traceability tools [6], and the system fails without reliable identification.

1.2. Plant Identification as Prerequisite for Proper Product Traceability

Genetic and sanitary characterization is the first step to build a detailed and useful information package; however, it is also essential to define all the other relevant input and output factors in order to trace the plants. Even if traceability in agriculture was clearly defined by the International Standard Organization (ISO) as "…the ability to trace and follow a food, feed, food producing animal or ingredients, through all stages of production and distribution" (ISO 8402:1994), the regulation is not completely clear: the United Nations General Assembly [8], United States Department of Agriculture [9] and European Union (2002/178/EC) have proposed various strategies, allowing specific regulations in member states, as in the EU. In this context several manufacturers, retailers, and service companies have already established or are establishing traceability procedures in agriculture with the primary aim of reducing business risks [6]. This voluntary approach to traceability can be standardized, following international standards such as the "Food safety management systems - Requirements for any organization in the food chain" (ISO 22000/2005), with the aim of harmonizing standards concerning food safety and Hazard Analysis and Critical Control Points (HACCP). Even if this approach is not obligatory and is more oriented to foods than plants, it represents a point of reference for stakeholders, supporting a virtuous circle of trust: the necessary information for standardizing production is linked to the identification of suppliers, participants in the production line, historic data and client feedback. All these data have to converge on the product in itself, possibly through an archiving and management platform.

With regard to plants as agricultural products, an information package has to be in place from an early stage, and relative information has to "grow" together with the plant: in some cases, this process is regulated by specific laws. For example, in the EU grapevines in the certificate category must be in line with the most recent directive, e.g., 2005/43/CE, and associated labels have to report essential data such as the nursery where they were produced. Thus, a basic level of information is already accessible for intermediate or final users: but is it enough? Surely food safety and plant health are guaranteed by the regulations, but some advantages can emerge in terms of additional value for the plant. In fact non-obligatory information about plant production can provide useful input for the farmer (for example a historical memory of the production process) or the final user, increasing his trust in the product.

This approach can be extended not only to plants which make up food chains, but to any kind of high value plant (e.g., ornamental plants [10]) for which production and marketing steps can be valorized. In this context, many propagative plants interested by certification disposals, are the ideal candidate for the application of a more-in-depth identification system and, consequentially, to be traced throughout production line and markets.

Plant traceability can be supported by information technology (IT), as it is for foods. The IT revolution, exemplified by the Internet and underlying information-technology hardware (e.g., increased computer processor speeds, increased data-storage capacity, electronic data capture and measurement devices) has made traceability and development of logistics management and monitoring economically feasible and enabled traceability of food products to survive the labyrinth of the agricultural product supply chain [11].

2. Plants with Implanted Microchips

2.1. The RFID Technology

RFID (radio-frequency identification) technology represents one of the most interesting tools in industrial and logistic sectors. In fact, it represents a new system to increase process efficiency in supply chains. The extremely rapid development of RFID in recent years is due to many factors, but a great impulse came from companies such as Wal-Mart, Metro, Tesco and government agencies such as the United States Department of Defence which has implemented RFID systems, thus stimulating global interested in what RFID can do and where it can be used [12]. Generally, RFID systems create a web used by companies to retrieve goods along a global supply chain. Companies can put various procedures into action on this "internet of things", even in multitasking [13].

The aim of RFID technology is to acquire information about objects, animals or people through microprocessors associated with them, as widely described with regard to its basic characteristics [12] [14] [15]. Generally, an RFID system is composed of an electronic label—a microprocessor/antenna system, generally called tag—a reader and a management system. The tag is includes a microprocessor control, equipped with memory, connected to an antenna and placed in a specific container. The tag incorporates a unique code readable by the reader. Tags are categorized by the type of energy source required for their operation. They are termed passive if they do not require a battery to operate, semi-passive if they have a battery to power the microchip and active if they are, together with the receiver and transmitter, battery powered. The reader interrogates and receives information from the tag and is connected to a computer system for management. Using radio waves, the RFID reader identifies a single item, and the process of communication between tag and reader is without physical contact. Compared to traditional systems of identification, such as barcodes, RFID offers many advantages, for example the possibility of reading a non-visible tag (i.e. when it is concealed inside packages or products) or simultaneous reading of multiple tags. Moreover, it is possible to store data within the tag memory or associate information using a specific database for storing the history of the product.

RFID is fundamentally based on wireless communication, using radio waves which form part of the electromagnetic spectrum, and it operates in unlicensed spectrum space, sometimes referred to as ISM (Industrial, Scientific and Medical) however the exact frequencies that constitute ISM may vary depending on the regulations in different countries [16]. These operating frequencies are generally considered to be organized into four main frequencies: Low Frequencies (LF, 30-300 kHz), High Frequencies (HF, 3-30MHz), Ultra High Frequencies (UHF, 300MHz-3GHz) and Microwave (2-30GHz). A typical RFID frequency for the LF band is between 120 and 145 kHz and it was the first frequency used, and it is still in use today. The HF band commonly focuses on a frequency of 13.56 MHz and, because of its reach, it is a worldwide reference. The UHF band has different frequencies, between 865 and 956 MHz [15]. When choosing a tag for an RFID system it is necessary to take into account various aspects, including size and shape, duration, resistance to external physical and chemical factors, orientation and distance from the reader, malfunctions near metals and liquids, fulfilment of local regulations, and memory capacity.

RFID represents an emerging technology that can be successfully applied to the management of supplies, production systems and logistics, although it is characterized by a potentially even wider range of applications. In fact, its applicative potential is underlined by a growing number of companies that have developed, either as test applications or for common usage, RFID, making this technology the most rapidly developing in terms of new potential applications for future business [17]. With the regard to the agricultural sector, initial applications were mainly directed toward product logistics [18] [19] [20] [21] [22], and the tagging of living organisms was relative to livestock. This sector represents one of the major beneficiaries of RFID applications in agriculture [23] [24] [25] [26] [27], also thanks to experiences gained in the veterinary sector on pets [28].

2.2. Tagging Plants with RFID Microchips

RFID tags can be associated with plants externally, also using electronic barcode systems which have been specifically developed [29]. A system has been proposed that uses using cell phones and portable computers to read RFID tags to which can be linked phenotypic, environmental and genomic data, thus creating an inexpensive system capable of recording and retrieving data from plant samples. The results suggested that RFID can be used in functional genomic screenings to record information involved in plant development or disease monitoring. Moreover, the proposed tags, Picotags (Inside Contactless, France) and Ti-tags (Texas Instruments, USA), maintained their features under different environmental conditions, ranging from -80 °C to 100 °C, as well as immersion in liquid nitrogen and various liquid solutions and could be autoclaved. These particular characteristics permit RFID systems to be used adequately in samples collected from plants.

Considering the uses of data linked to tags, there are no conceptual differences between external tagging or implanting tags inside plants: a band containing a microchip that is attached to a tree can provide the same information as an implanted one [30], however there are some issues that should be considered. External tagging is simple, inexpensive and can be performed at any stage of plant growth, yet an external tag can be removed or damaged more easily, making this approach optimal for sample tracing [29], in a context in which plants are not commercialized or permanent identification is not necessary. Otherwise, internal insertion eliminates the risk of loss without damaging the plant itself, satisfying the permanent identification pre-requisite for traceability of commercialized plants.

Keeping in mind plant histology and organ size, different techniques and tag positions have been proposed for implanting microchips, although it does not seem possible to standardize internal RFID tagging in plants. For large caliper trees such as adult *Cypress* spp. [31], the insertion of tags after trunk drilling did not require particular attention with regard to position or extension of drilling, thanks to the fully developed anatomy of the trees. Damage to plants was not reported and tags working at 131.6 kHz were used to memorize an identification number associated to informational sheets stored in a digital database. The proposed system responded perfectly to needs for health status monitoring of adult trees which are important for the community or are considered national monuments, as was the case of the cited *Cypress* spp. or *Cactus* spp. [32].

When dealing with small trees, such as those typically found in tree nurseries, tag implantation inside organs of small size requires specific methods. A report regarding microchip implantation in *Citrus* spp. at nursery stage [33] is an example of a plant tagging procedure that was reliable, durable and secure. The described method involved an upright T-cut above the graft union during active tree growth, followed by an insertion procedure that was similar to that for budding of citrus nursery trees. This method for implanting microchips was specifically developed for smaller caliper trees and could also be applicable to other woody plant species. Commercial microchips such as AVID Code 14 mm, AVID Euro 12 mm, and ADS 12 mm were used to assign unique identity to citrus trees, and tests of readability through wood were performed in order to evaluate the applicability of this approach to plants. It was reported that the signal penetration varied significantly depending on the scanning devices and, in a more limited way, depending on the wood type. In any case, considering plant growth, the reading of microchips can be assured in most woody plant species for 10 years or more when appropriate RFID scanners are selected.

To implant tags inside grapevine plants at nursery stage, the preferred place for microchip localization is in the pith [34]. Tags were inserted inside the pith of rootstocks, following two different, specifically designed procedures. The first involved microchip insertion after direct drilling of the pith from the distal cut of the rootstock just before grafting, followed by microchip localization below the grafting point, while the second procedure was carried out after grafting and consisted in a "U" cut performed laterally on the rootstock below the grafting point, involving tissues from bark to pith. In this latter procedure, the microchip was thn located inside the pith, and cut tissues were manually reassembled. Transponder glass tags were used (2.1 mm in diameter and 12 mm in length) which worked at a frequency of 125 KHz.

In order to evaluate the effects of microchip implantation, growth parameters and histological observations have been examined for small caliper plants. Generally, the available methods for microchip implantation in small caliper plants have not been found to compromise growth [33] [34]. The most invasive implantation procedure caused a reduction in branch development in grapevine after the subsequent year of growth, but plants can reach normal development from the second year after implantation [35]. Histological observations of plants with implanted RFID microchips have been reported for grapevine [35]: the suggested procedure caused no effects on tissue development or, in the most invasive method, the functionality of xylem was partially lost. However, these effects seem limited to the first months after implantation, after which time the newly formed xylem restore their transversal continuity. Results obtained in small caliper trees implanted with RFID tags suggest that tagging procedures can be safe if adequately performed.

Commercialized RFID-tagged plants have been reported for a *Platanus* variety resistant to canker stain, with the authenticity of each tree guaranteed by a microchip in the trunk [36].

3. How RFID Can Support Farm Management

3.1. RFID for Information Management Purposes in Agriculture

Agricultural management systems involving RFID have been proposed in livestock [37] [38] [39] [40] [41] in which various animal parameters, input and output can be checked. RFID systems permit remote monitoring of animals using dedicated hardware and software. Information related to the organism, item or environmental factor is accessible via various monitoring stations and operators, and can be updated and shared, representing a new approach to farm management.

Recently an integration of information systems has been proposed to advise managers of formal instructions, recommended guidelines and documentation requirements for various decision making processes [42]. This is the case of the EU-funded project FutureFarm, in which a new model and prototype of a Farm Information Management System (FMIS), which meets these changing requirements, will be developed. Among possible implementations of the component-based model of the FMIS, sensor readings from process activities involving RFID and Global Positioning System (GPS)-based technologies such as wireless systems for data transfer have been suggested. In this approach a database containing information on the operation history of the farm, (such as a local database on a farm's pc or the centralized one proposed by the Danish Agricultural Advisory Service [43]), can be usefully integrated in systems like FMIS.

Nowadays farm planning has to take into account the dynamic interactions between machine, biology and meteorological conditions [44]. The need for better analysis and transformation of the collected data calls for better precision and integration of planning and farm control functions, thus recreating computer-based production as found in the industrial sector and embracing customized production followed by dynamic operation planning and control [45]. This approach has been successfully adopted in industry, and it has been shown how effective an integrated control of work operations can be when based on on-line measurements combined with a database and decision support information [46] [47]. In this context, supply chain activities have found benefit from integrating IT and information systems [48], and this approach can also be transferred to the agricultural sector. As proposed, a decisive prerequisite for the development of comprehensive and effective ICT-system supporting the task management efforts is represented by a detailed structuring and formalization of physical entities and the information which surrounds the planning and control of farm operations using efficient mobile working units in automated agricultural plant production systems [42]. In this context, RFID technology could represent a useful tool for linking the physical entities (that is, the tagged plant) to the farm system.

In order for RFID-tagged plants to be items in a digital information management system, computer storage databases have been proposed. To manage samples for plant pathology or biology purposes, a labeling system to associate a RFID tag to a MMS message was suggested [29]. The GPS reading, date, and additional data were entered on a cell phone, the receiver was switched to writing mode and the information was recorded on the RFID tag. Using this approach, photographs of the plant were taken and audio information about the experiment directly linked to the picture. The name of the photo was then changed to the RFID identifier and entered into the cell phone contact database, sent to a personal email account via cell phone MMS with the RFID number as the subject line for the email message. A system to transfer the RFID data to an external database was developed, transferring information to portable computers via Bluetooth, IR, or USB connections. The information was then exported into Filemaker (Santa Clara, USA) databases that were used on Macintosh and PC Windows platforms.

To connect a wider range of stakeholder involved into a plant production line, online databases were proposed. A specifically developed database has been designed for tracking RFID tagged grapevines from the nursery stage [49]. The proposed system permits access, by digitally entering RFID codes into the search field of the database, to datasheets relative to tagged plants with all protagonists of the production line registered. Datasheet can be considered a sort of electronic identity card (eID) for each individual plant, as opposed to traditional labels placed on groups of grapevines joined by laces. The various users of the database have different duties and privileges. The nursery stock is controlled by the administrator of the database who can edit clone and rootstock files, and match them using the tag code. Thus, the code (linked to one single microchip) is used to edit the eID of each plant subsequently produced by the nursery. Grapevine nurseries that access the database can view all data regarding clones and rootstock, and can add commercial notes relevant to the grafted cuttings to the eID. The grapevine grower, in addition to being able to consult cutting files, can use the tag code to access the eID of purchased plants, edit personal data fields, and manage entire groups of marked plants (the so-called "virtual vineyards").

Online database systems using RFID codes associated to plants can also be used to retrieve information about relative wines [50]: interestingly, this application could be combined with corks containing an RFID inlay for high value wines [51] [52].

3.2. RFID/GIS Applications: Outlook in Precision Farming

A wireless sensor network is a system made up of radio-frequency transceivers, sensors, microcontrollers and power sources and its applications in agriculture and the food industry are still rare [53]. These systems offer the advantages of reducing wiring costs, effectively monitoring the environment, they are small in size and the cost of sensors is limited. Moreover, they can be implemented in precision agriculture [53].

Wireless technology has been developed for spatial data collection for crop management and spatial-variability studies [54] [55]. In addition, for manual fresh fruit harvesting RFID systems were integrated for yield mapping procedures to overcome the limitations of existing systems [56]. In this latter case a novel commercial location technology based on RFID was proposed to establish each tree's geo-location and the associations between trees and the bins involved in their harvesting. RFID tags were attached to both individual trees and bins used for harvesting. The proposed method represents a useful solution combining RFID technology and precision farming, even if some concerns arose from the RFID reader miss ratio for detection of the bins. Thus, attachment of RFID tags on suitable tree branches to achieve 100% detection needs enhancement, and additional trials need to be performed to determine the best positions and orientations of tree tags for the RFID reader to successfully detect them [56]. Even if the proposed method is not yet optimized (mainly due to technological limitations), an integrated farm system via RFID can represent a significant improvement for harvesting all kinds of woody plants: the link between tagged plant, bins and storehouse can help the farmer in management and product logistics.

The virtualization of vineyards by combining GPS technology and computer-aided design (CAD) is another interesting possibility: it is currently possible to record geographic profiles of fields such as vineyards easily and use them in CAD software to draw structural elements [57]. In the urban context, geospatial tools such as GPS and the Geographical Information System (GIS) can provide timely and extensive spatial data to arrive at plant attributes that can be adapted for applications including data fusion, virtual reality, three-dimensional visualization, internet delivery, and modeling [58] [59]. This approach to urban green management or forestry monitoring can be extended to orchards or vineyards where plants are tagged with RFID: geo-referred data from transplant machines can be matched to RFID labels on individual plants, headland of vineyards or wine bottle [60]. Potentially, subjects containing a microchip will be identified by a code, associated with the microchip itself, and they will be located with GIS on a three-dimensional electronic map, recreating a virtual vineyard or orchard [49]. Electronic mapping techniques could be implemented in indoor areas (i.e. screenhouses for grapevine collections, as reported in clonal selection procedures for grapevine) and outdoor areas, such as nurseries and vineyards, considering distinctive and unique properties of these different environments. The advantages of this approach are linked to the possibility of remotely monitoring vineyards, filing and managing useful data associated with the plants (e.g. identity, sanitary status, certification, and cultural practices particularly using technical and plant health files), and having a durable, safe and detailed vineyard information map [49].

Mobile devices like netbooks, tablet-PCs or smartphones can represent optimal instruments for consulting and updating the virtual orchard or vineyard from the field, while time-consuming operations can be carried out using a desktop device. This strategy has been described for contextualized vineyard management [61] where tags were placed in the field and decoded by mobile devices such as mobile phones or PDAs. In the work, a geo-referring was performed which automatically associated a tag to a field location in the relevant database tables or records and also permitted access to contextual information or services. The proposed system can be considered user-friendly: by pointing a mobile device at a tag, the farmer can in a simple way download information such as climatic data or upload information such as disease and pest incidence. The possibility of performing these operations from the field without having to provide coordinates or any other references avoids having to return to a central office for retrieving information or instruments [61]. Even if this system is not directly linked to plants, it represents an interesting approach in integrating RFID in precision viticulture, and the availability of microchip implantation methods for grapevine offers further options in this sector.

4. CONCLUSIONS

Quality certification schemes, such as 2006/510/EC, do not provide for implementation of RFID microchips, and they do not enhance the intrinsic value of a product such as a plant and its derivatives. Experience with livestock suggests that traceability systems such as RFID can represent a useful, or even essential resource for agricultural product safety. While plants cause less concern about human health when compared to the meat production chain, implications do exist in terms of the environmental impact of production systems, the worldwide spread of plant pathogens (in particular viruses) and chemical residuals. Plant traceability by RFID can be a useful tool when it comes to risk management.

When considering agricultural products such as plants, quality is also made up of intangible elements that consumers cannot verify themselves [62]: beyond respect for the environment or equity in transactions, genetic identity and health status are not directly distinguishable, and consumers have to trust the information provided. Thus quality becomes objectified and symbolized by various signs which form the basis of both consumer recognition and market valorization [63]. In this context, identification and traceability when guaranteed by RFID systems can enhance this aspect of product quality. Moreover, marketing strategies can obtain benefits in product differentiation relative to tagged plants. Currently firms are able to evade price-based competition between identical products while also responding to the increasingly differentiated demand structure associated with a postFordist global economy, adding value to products to achieve an above-market price that economists refer to as a 'differential rent' [63].

Also, farm management systems can find benefits with RFID. In fact, a farm information management system such as FMIS focuses on the farm manager and the everyday management problems, identifying central entities in the proposed system that includes the farm manager, the fields, the products and production input [42]. In this context RFID-tagged plants can be useful for linking the core of plant production to the information management system. The main benefits are not only associated with improving consumer trust in certificated plants, but also tagged plants can represent the heart of nurseries and farms, where digitalized management is an essential tool for facing future challenges.

In the near future some features of RFID-tagged plants might be developed in light of new Internet technology, namely Web 2.0, with increased functionality of Web sites [64]. In fact, proposed systems for information management of farms and database storage systems for tagged plants have elements in common with Web 2.0, and new Web-based tools can be integrated to share information of tagged plants and their relative input/output. Moreover, network- or Internet-friendly equipment such as RFID microchips associated with organisms or samples represents an optimal link between laboratories; Web-based systems have been reported in livestock [41] and grapevine farming [49], providing advantages for researchers. The widespread use of mobile devices for Web access can support Web-based resources such as online databases for farm management or field monitoring. When combined with RFID tags on plants and materials like bins, this approach can create an "Internet of things" in plant farming as well.

The cost of RFID systems for plants may represent a limitation for the spread of this technology in plant identification and traceability. However the increase in sales of microchips reveals a general interest in using this identification system in various fields. As reported by authoritative market research [65] [66], 2.4 billion tags were sold from 1946 to 2005: the trend is increasing constantly, with 600 million tags in 2006, 1.97 billion in 2008 and 2.35 billion in 2009. With regard to agriculture, the number of microchips used for livestock in 2009 was about 105 million. Growing markets are reflected in the price of tags, mainly due to reduced assembly costs. Considering one of the simplest and most frequently used tag (class 0 or 1 passive microchip), in 2004 the cost was € 0.47 and it is predicted to be € 0.26 in 2012, largely because of a reduction in the cost of materials [67]. With the aim of implementing RFID systems in each single plant, the current cost of tags may represent the main limitation. However, in light of the high value of plants such as woody perennials, the cost can seem affordable even now.

The increasing demand for traceability, which requires not only rigorous inspections but also systematic detection, is focused on labeling and the recording of quality and safety parameters while keeping track of the entire agri-food production chain, from farms to the consumer's table [53]. RFID has been termed the most important identification tool to establish an effective traceability system [68] and even if microchip implantation procedures in plants have recently been developed, tagged plants and their digital management most probably will have significant impact on farms in the near future.

ACKNOWLEDGMENTS

The authors would like to thank Prof. Enrico Triolo, Prof. Enrico Rinaldelli, Dr. Roberto Bandinelli and Mrs. Barbara Gini for discussions which helped to develop ideas put forward.

REFERENCES

[1] Thysen, I. (2000). Agriculture in the information society. *J. Agric. Eng. Res.,* 76, 297-303.

[2] Mink, G.I. (1998). Virus certification of deciduous fruit trees in the United States and Canada. In A. Hadidi, R.K. Khetarpal, H. Koganezawa, *Plant Virus Disease Control* (pp. 294-300). St. Paul, USA: APS Press.

[3] Sarig, Y. (2003). Traceability of food products. *CIGR Journal of Scientific Research and Developments,* 5(12), 54-65.

[4] European Commission (2000). The European White Paper on Food Safety [online]. Available from: http://ec.europa.eu/dgs/health_consumer/library/pub/pub06_en.pdf.

[5] NACA/FAO (2000). The Bangkok Declaration and Strategy on Aquaculture Development [online]. Available form: http://www.oceansatlas.org.

[6] Regattieri, A., Gamberi, M., Manzini, R. (2007). Traceability of food products: General framework and experimental evidence. *Journal of Food Engineering* 81, 347-356.

[7] Luvisi, A. (2007). Radiofrequency technology (RFID) in grapevine nursery for sector traceability. In: *Proceedings of ID WORLD International Congress 2007,* Milano, 26-28 novembre 2007.

[8] United Nations, (1986). *UN Guidelines for Consumer Protection.* General Assembly resolution, 39/248.

[9] United States Department of Agriculture, (2002). *Traceability for Food Marketing & Food Safety: What's the Next Step? Agricultural Outlook*, January–February 2002, 21–25.

[10] Triolo, E., Panattoni, A., Pagano, M., Bandinelli, R., Rinaldelli, E., Luvisi, A. (2010). *Rose e microchip: una nuova passione?* Agronotizie, 8 giugno 2010.

[11] Buhr, L.B. 2003. Traceability and Information Technology in the Meat Supply Chain: Implications for Firm Organization and Market Structure. *Journal of Food Distribution Research,* 34(3), 13-26

[12] Ngai, E.W.T., Moon, K.K.L., Riggins, F.J., Yi, C.Y. (2008). RFID research: an academic literature review (1995-2005) and future research directions. *Int. J. Production Economics,* 112, 510-520.

[13] Violino, B. (2005). Leveraging the Internet of things. *RFID Journal,* November/December, 1-2.

[14] Finkenzeller, K. (2003). *RFID handbook: fundamental and applications in contactless smart cards and identification* (2nd edition). Munich, D: Wiley.

[15] Ward, M., van Kranenburg, R. (2006). RFID: Frequency, standards, adoption and innovation. *JISC Technology and Standards Watch,* May 2006, 4-15.

[16] Brock, D. (2001). *The Physical Markup Language [online].* Available from: http://xml.coverpages.org/PML-MIT-AUTOID-WH-003.pdf

[17] Heinrich, C. (2005). *RFID and beyond.* Indianapolis, USA: Wiley.

[18] Amador, C., Emond, J.P., do Nascimento Nunes, M.C. (2009). Application of rfid technologies in the temperature mapping of the pineapple supply chain. *Sensing and Instrumentation for Food Quality and Safety,* 3(1), 26-33.

[19] Jones, P., Clarke-Hill, C., Comfort, D., Hillier, D., Shears, P., (2005). *Radio frequency identification and food retailing in the UK. British Food Journal*, 107(6), 356-360.

[20] Purvis, A.C., McGlasson, W.B., Kanlayanarat, S. (2006). Progressive challenges in horticultural supply chains: some future challenges. *Acta Horticulture*, 712(1), 39-49.

[21] Gandino, F., Montrucchio, B., Rebaudengo, M., Sanchez, E.R. (2009). On improving automation by integrating rfid in the traceability management of the agri-food sector. *IEEE Transactions on Industrial Electronics,* 56(7), 2357-2365.

[22] Jedermann, R., Ruiz-Garcia, L., Lang, W. (2009). Spatial temperature profiling by semi-passive rfid loggers for perishable food transportation. *Computers and Electronics in Agriculture,* 65(2), 145-154.

[23] Buguk, C., Ervin, R.T., Eberspacher, J. (1998). Economic analysis of an alternative cattle identification system used to decrease hide damage. *Journal of the American Leather Chemists Association*, 93, 248-254.

[24] Caceci, T., Smith, S.A., Toth, T.E., Duncan, R.B., Walker, S.C. (1999). Identification of individual prawns with implanted microchip transponders. *Aquaculture,* 180, 41-51.

[25] Hsu, Y.C., Chen, A.P., Wang, C.H. (2008). A rfid-enabled traceability system for the supply chain of live fish. In: *Proceedings of IEEE International Conference on Automation and Logistics, ICAL 2008, September 1 - September 3, 81-86.*

[26] Reiners, K., Hegger, A., Hessel, H.F., Bock, S., Wendl, G., Van den Weghe, H.F.A. (2009). *Application of rfid technology using passive hf transponders for the individual identification of weaned piglets at the feed trough. Computers and Electronics in Agriculture*, 68(2), 178-184

[27] Shanahan, C., Kernan, B., Ayalew, G., McDonnell, K., Butler, F., Ward, S. (2009). A framework for beef traceability from farm to slaughter using global standards: An irish perspective. *Computers and Electronicsin Agriculture,* 66(1), 62-69.

[28] Sorenson, M.A., Buss, M.S., Tyler, J.W. (1995). Accuracy of microchip identification on dogs and cats. *Journal of the American Veterinary Medical Association,* 207, 766-767.

[29] Kumagai, M.H., Miller, P. (2006). Development of electronic barcodes for use in plant pathology and functional genomica. *Plant Molecular Biology,* 61, 515-523.

[30] Grieco, P.D., Mendoliera, S., Castoro, V., Vitelli, V., Cellini, F., Agnello, A., Buccigrossi, F., Vigo, G. (2006). La tecnologia RFID per la tracciabilità e la certificazione delle produzioni vivaistiche. *Rivista di Frutticoltura*, 10, 60-70.

[31] Battezzati, L., Miragliotta, G., Perego, A. (2006). RFID alla prova dei fatti [online]. Available from http://www.rdlog.it/doc/Report_RFId_2006.pdf

[32] Associated Press (2008). *Theft deterrence for an Arizona icon.* New York Times, October 12 2008, 39.

[33] Bowman, K.D. (2005). Identification of woody plants with implanted microchips. *HortTechnology,* 15(2), 352-354.

[34] Bandinelli, R., Triolo, E., Luvisi, A., Pagano, M., Gini, B., Rinaldelli E. (2009). Employment of radiofrequency technology (RFId) in grapevine nursery traceability. *Advances in Horticultural Science* ,23(2), 75-80.

[35] Luvisi, A., Panattoni, A., Bandinelli, R., Rinaldelli, E., Pagano, M., Gini, B., Triolo, E., (2010). RFID microchip internal implants: Effects on grapevine histology. *Scientia Horticulturae,* 124, 349-353.

[36] INRA (2008). Platanor® Vallis clausa, a plane tree variety resistant to canker stain [online]. Available from http://www.international.inra.fr /press/platanor_r_vallis_clausa

[37] Haapala, H.E.S. (2003). Operation of electronic identification of cattle in Finland. In: *Proceedings of the 4th European Conference in Precision Agriculture,* Berlin, Germany, June 14–19.

[38] Kononoff, P.J., Lehman, H.A., Heinrichs, A.J. (2002). A comparison of methods used to measure eating and ruminating activity in confined dairy cattle. *J. Dairy Sci.,* 85, 1801-1803.

[39] Nagl, L., Schmitz, R.,Warren, S., Hildreth, T.S., Erickson, H., Andresen, D. (2003). Wearable sensor system for wireless state-of-health determination in cattle. In: *Proceedings of the 25th IEEE EMBS Conference,* Cancun, Mexico, September 17–21.

[40] Taylor, K., Mayer, K., (2004). TinyDB by remote. In: *Proceedings of Australian Mote Users' Workshop,* Sydney, Australia, February 27.

[41] Voulodimos, A.S., Patrikakis, C.Z., Sideridis, A.B., Ntafis, V.A., Xylouri, E.M. (2010). A complete farm management system based on animal identification using RFID technology. *Computers and Electronics in Agriculture,* 70, 380-388.

[42] Sørensen, C.G., Fountasb, S., Nashf, E., Pesonend, L., Bochtisa, D., Pedersene, S.M., Bassoc, B., Blackmoreg, S.B. (2010). Conceptual model of a future farm management information system. *Computers and Electronics in Agriculture,* 72, 37-47.

[43] DAAS (2009). *Danish Field Database.* The Danish Agricultural Advisory Service.

[44] Kuhlman, F., Brodersen, C. (2001). Information technology and farm management: developments and perspectives. *Computers and Electronics in Agriculture,* 30, 71-83.

[45] Nagalingam, V. Ser, Grier, C.I. Lin (2008). CIM – still the solution for manufacturing industry". *Robotics and Computer-Integrated Manufacturing*, 24(3), 332-344.

[46] McCarthy, J.J. (1990). *The challenge of CIM in the process industries. ISA Transactions,* 29(1), 53-56.

[47] Riezebos, J., Klingenberg, W., Hicks, C. (2009). Lean production and information technology: connection or contradiction? *Computers in Industry,* 60(4), 237-247.

[48] Gunasekaran, A., Ngai, E.W.T. (2004). Information systems in supply chain integration and management. *European Journal of Operational Research,* 159(2), 269-295.

[49] Luvisi, A., Triolo, E., Rinaldelli, E., Bandinelli, R., Pagano, M., Gini, B. (2010). Radiofrequency applications in grapevine: From vineyard to web. *Comput. Electron. Agric.,* 70, 256-259.

[50] Metapontum Agribios s.r.l. (2008). *Portale vitivinicolo [online].* Available from http://www.certabasilicata.it

[51] Collins, J. (2005). Wine Bottles Get Corked With RFID [online]. Available from http://www.rfidjournal.com/article/articleview/2117/1/1/

[52] Launois, A. (2008). RFID tracking system stores wine bottle data [online]. Available from http://www.foodproductiondaily.com/ Packaging/RFID-tracking-system-stores-wine-bottle-data

[53] Wang, N, Zhang, N., Wang, M. (2006). Wireless sensors in agriculture and food industry-Recent development and future perspective. *Computers and Electronics in Agriculture*, 50, 1-14.

[54] Gomide, R.L., Inamasu, R.Y., Queiroz, D.M., Mantovani, E.C., Santos, W.F. (2001). An automatic data acquisition and control mobile laboratory network for crop production systems data management and spatial variability studies in the Brazilian center-west region. In: *The American Society of Agriculture Engineers,* St. Joseph, Michigan, USA, ASAE Paper No. 01-1046.

[55] Mahan, J., Wanjura, D. (2004). Upchurch, Design and Construction of aWireless Infrared Thermometry System. In: *The USDA Annual Report,* May 01, 2001–September 30, 2004, Project Number 6208-21000-012-03.

[56] Ampatzidis, Y.G., Vougioukas, S.G., Bochtis, D.D., Tsatsarelis, C.A. (2009). A yield mapping system for hand-harvested fruits based on RFID and GPS location technologies: field testing. *Precision Agriculture*, 10, 63-72.

[57] Pergher, G., Vieri, M. (2007). Progressi della ricerca nell'ingegneria delle produzioni viticole. In: Proceedings of Convegno Nazionale "Tecnologie innovative nelle filiere: orticola, vitivinicola e olivicola-olearia", *Pisa e Volterra,* 5-7 settembre 2007, 229.

[58] Ward, K.T., Johnso,n G.R. (2007). *Geospatial Methods Provide Timely And Comprehensive Urban Forest Information.* Urban Forestry & Urban Greening, 6, 15-22.

[59] Wu, C., Xiao, G., McPherson, G.E. (2008). *A method for locating tree-planting sites in urban areas: A case study of Los Angeles*, USA. Urban Forestry & Urban Greening, 7(2), 65-76.

[60] Vieri, M. (2007). Dispositivi e procedure nella viticoltura di precisione al fine della tracciabilità di prodotto e della ecocompatibilità di processo. In: *Proceedings of L'e-nell'ingegneria agraria, forestale e dell'industria agro-alimentare*, Firenze 25 - 26 ottobre 2007.

[61] Cunha, C.R., Peres, E., Morais, R., Oliveira, A.A., Matos, S.G., Fernandes, M.A., Ferreira, P.J.S.G., Reis, M.J.C.S. (2010). The use of mobile devices with multi-tag technologies for an overall contextualized vineyard management. *Comput. Electron. Agric.,* doi:10.1016/ j.compag.2010.05.007

[62] Renard, R.C. (2005). Quality certification, regulation and power in fair trade. *Journal of Rural Studies,* 21, 419-431.

[63] Valceschini, E., Nicolas, F. (1995). *La dynamique e´conomique de la qualite´ agro-alimentaire.* In E. Valceschini, F. Nicolas, Agro-alimentaire: une e´conomie de la qualite´ (pp. 15-38). Paris, France: INRA Economica.

[64] O'Reilly, T. (2005). What is Web2.0. Design patterns and business models for the next generation of software [online]. Available from http://www.oreillynet.com/pub/a/oreilly/tim/news/2005/09/30/what-is-web-20.html

[65] IDTechEx (2005). RFID tag sales in 2005-how many and where [online]. Available from: http://www.idtechex.com/research/ articles/rfid_tag_sales_in_2005_how_many_and_where_00000398.asp

[66] IDTechEx (2009). RFID Forecasts, Players and Opportunities 2009-2019 [online]. Available from: http://www.idtechex.com/ research/reports/rfid_forecasts_players_and_opportunities_2009_2019_ 000226.asp

[67] CNIPA (2007). *Linee guida per l'impiego dei sistemi RFId nella Pubblica Amministrazione*. I Quaderni, 30 febbraio 2007, 45-47.

[68] Sahin, E., Dallery, Y., Gershwin, S. (2002). Performance evaluation of a traceability system: an application to the radio frequency identification technology. In: *Proceedings of the 2002 IEEE International Conference on Systems,* Man and Cybernetics, vol. 3, Yasmine Hammamet, Tunisia, October 6–9, 647-650.

In: Radio Frequency Identification ISBN 978-1-61122-416-0
Editor: Alison R. McAdams, pp. 75-91

Chapter 4

RADIO FREQUENCY IDENTIFICATION FOR THE ORGANIZATION OF MEDICINE DISASTER AID

Christian Di Filippo
Centre Hospitalier Universitaire de Lyon, Lyon, France

INTRODUCTION

Support and transmission of medical information are still challenging the field of Disaster Medicine.

The department. of SAMU (Service emergency medical assistance) of Lyon (France) and its unit of Disaster Medicine are continuously looking for new technologies to enhance medical doctors actions and ensure traceability.

We therefore tried to apply the RFID method in this context, and to computerize the medical chain of relief at the scene of a disaster.

We will at first define the so-called situation of a disaster, and the role of French SAMU. Second, we will describe the management of these situations as planned by the Government in the red channel. We will then give details of the means and methods for computerization of these emergency resources, with particular attention to the use of RFID.

General Considerations

1. The Disaster Medicine in France

Definition

The Disaster Medicine is a branch of Emergency Medicine devoted to accidents or disasters involving mass casualties.

The disaster is defined by the inadequacy of relief needs and available resources (resources are overwhelmed because of exceptional uncontrolled events). This situation requires specific organization and management, different from those of regular emergency medicine.

The complex modern society carries on an increasing number of major risks, changing threats and new crises that must generate actions for prevention, and management of the collective risk.

Risks

The main risks are classically divided into 3 groups:

- Technological risks: directly related to human industrial activities and traffic large scale accidents
- Natural hazards: various disasters (storms, earthquakes, tsunami ...) epidemics.
- Sociological risk with disasters related to armed conflict or civil, and acts of terrorism.

Diverse combinations of these risks can also occur.

The type of risk may be either conventional (classic injuries) or NRBC (nuclear, radiological, biological, and chemical).

Response

The medical management of victims is part of the Public Health Program. To address these situations, organizing preventive relief is mandatory, in order to improve the quality of care to victims, save as many human lives as possible, and limit the sequelae of injuries.

Given the diversity of situations at major collective risk, the French governments have undertaken a large interdepartmental workup to implement the entire national territory with unified means.

The general emergency plans include the red plan which ensures rapid organization of means at the level pre-hospital support of victims, and the white plan issued to receive these victims in hospitals.

The red channel is established by Decree No. 88-622 of May 6, 1988. It organizes the procedures of emergency and how to engage in an event that could cause many casualties. The pre-hospital activated response aims to control the consequences of the event, i.e. how to allocate and organize means, and how to define the roles and responsibilities of each stakeholder. [1], [2], [3]

2. The SAMU and its Department of Disaster Medicine

SAMU (Service d'Aide Médicale Urgente)

The French SAMU organization is dedicated to emergency medical assistance, providing specialized human skills and track hospital equipment to bring severely ill and/or injured patients to the medical intensive care unit. It completes specific objectives of the French National Health Organization:

- Medical regulation of telephone calls, to adjust rescue means.
- Regulation of pre-hospital emergency means, to guide rescue crews (fire fighters and private ambulances) and mobile emergency crews to adequate hospital units
- Provide pre-hospital teams and mobile medical tools for resuscitation. (SMUR)
- Trigger alert and manage crisis in white-red plans. SAMU center plays a pivotal role to dispatch and balance means between the place of the disaster and in-hospital tools. On-site rescue is conducted by the so-called director of medical care (1st on the scene or more experienced), whose role is to organize medical emergency in association with the emergency physicians crew. They collaborate with rescue workers, fire fighters who insure safety on-site and delineate prohibited areas in case of contamination risk, hold injured persons, initiate decontamination, and transit victims throughout the chain of emergency. SAMU ensures the medicalization of each link in the chain and provides adequate equipment on-site. (drugs, resuscitation equipment, mobile communication tools)

The SAMU Department and its Disaster Medicine Unit

The Disaster Medicine Unit is in charge of:

- Maintenance and management of hardware dedicated to disaster situations.
- Training of doctors through a specific academic Program Training in disaster medicine.
- Organization of paramedics training in practice by simulating disaster situations
- Research for new intervention strategies and materials.

The primary objective is to improve the efficacy and results of care in the context of the Red Plan.

We recently developed an automated telephone reminder service to recruit medical staff in case of emergency and red channel activation.

Improve practice of Emergency Medicine needs to be alerted of any new technologies that could be applied to Disaster Medicine.

In this field, the computerization of the chain of emergency appears to be of high interest. The goal is to shorten time to intervention and use equipment to backup any medical information. RFID has emerged as a promising technology for this purpose.

3. RFID Technology and its Current Use [4], [5], [6], [7]

RFID is a method to store and retrieve data by using markers called RFID tag. These tags include an antenna coupled to an electronic chip that enables them to receive and respond to radio queries issued from a transceiver.

These tags can be embedded in objects and even implanted in bodies. Similarly to the bar codes, the RFID tags fall into the category of automatic identification technologies. The barcode is a pioneer technology in the field of identification, but does not allow storage of large amounts of data, whereas the RFID tag can store up to 10 kbit data.

The RFID tag is a miniaturized system, of one millimeter in size. It consists of an antenna (a coil of copper plate) and a microchip the size of a pinhead. It must be able to both store informations and transmit data.

The so-called passive RFID tags use the lector short-distance energy. These are the most commonly used. The lector is the active device which feeds the labels present in the field by a process known as "TV supply". This mechanism can produce energy (antenna coiling through an electromagnetic field), which provides batteries for virtually unlimited use.

The labels utilize the electromagnetic field (created by the lector), which activates the chip; the latter will then run the programs it was designed to. The transmission is ensured through modulation of amplitude or phase of the carrier frequency. The reader receives the informations and converts them into binary code. Reversely, the lector can send out informations by modulating the carrier. The modulations are analyzed by the chip and digitized.

Of note, labels can also be differentiated by the type of utilized memory. The most commonly used is an EEPROM (Electrically Erasable Programmable Read Only Memory) which is erasable and programmable. An EEPROM memory allows up to 500,000 rewrites. The type of memory allows to define the reading and writing modes.

Industry is the main purpose of this growing technology, which aims to increase sales, reduce inventory and inspection time, establish permanent inventories of products, target withdrawal procedures, and to provide consumers with detailed information about the product. The long distance capture of informations enables real-time monitoring of freight flows. RFID continues to develop in many areas. It enables to warrant identification, tracing, securing in various activities.

RFID is also used for identification of individuals, by storing biometric information and digital photos of people on passports or identity cards.

RFID is already involved in the authentication and traceability of imported drugs and also helps in the management of some pharmaceutical industries. [8], [9] Conversely, the concept of medical record cards is on its way.

Implementing RFID in a triage system during a simulated mass casualty situation has already been tested in military. [10]

The use of RFID technology to enhance hospital response to mass casualty events has already been studied. [11]. The objective is to improve traceability of patient medical information.

Authors stress the importance of traceability of medical information in disaster situation. [12], [13], [14], [15], [16] They have tested in the hospital, computer systems and barcodes for the media in medical information. The results of these studies seem promising.

The following text will give details on RFID application in the field of Disaster Medicine in prehospital setting.

Materials and Methods

1. The Red Channel and the Current Channel of Relief

General Organization

Each red plan is prepared under the departmental authority of the prefect, in conjunction with local authorities and the services and agencies involved in emergency medical assistance and medical transportation.(SAMU and fire services and rescue) The warden triggers the plan whenever regular means are overwhelmed; in this context, some state services will need to be serially activated. The SAMU physician in charge of calls regulation is informed at first and triggers the plan.

Throughout the red channel period, any relief is placed under the authority of the prefect. On-site command is headed by both the commander of relief operations (departmental director of fire services and rescue or his representative) and the director of medical care (1 SAMU doctor on site later replaced by a SAMU physician experienced in managing such crisis).

- Since the outbreak of the plan, a command post is settled in the operation room of the Prefecture. An operational command post settles near the site of intervention, and ensures coordination and implementation of emergency resources.
- The commander of relief operations guarantees the security on-site and delineate exclusion zones and areas for deployment of medical aid.
- SAMU team promptly brings on-site human resources (emergency physicians, intensivists, nurses, and paramedics), technical equipment for resuscitation (Mobile Post Health), communication tools (walkie talkie, satellite phone), medical records, command vehicle in permanent contact with the dispatch center for patients hospital referral. The medical director is in charge of organizing and managing the medical deployment.

The medical-chain starts with the firefighters team gathering victims and bringing them to the sorting center, where a physician categorizes cases in absolute or relative emergency. Each victim is given a medical card that provides "traceability" from release to final destination.

The victims then enter in the advanced medical post, receive medical care and are settled for transport. The identification of the victim, the level of emergency, diagnosis, therapeutic and manuscripts are recorded on the medical cardboard which is hung around the neck of the patient.

Last link is the sorting center, informed by the vehicle's command SAMU about the patient destination. Evacuations are shared by fire services and rescue, SMUR and licensed private health carriers. The medical card is all along following the victim; traceability therefore relies upon this document.

Informations are Stored through

- The Medical-sheet cardboard which represents the medical record and ensures the traceability of the patient data to the hospital. Each card carries a unique number. It contains tables filled by each physician, and displaying diagnosis and therapeutics.
- Radio-communications between doctors and command vehicle where status of each victim is transcribed manually.
- Telephone link between command vehicle and the dispatch center to guide patients to hospital.
- Radio-communication between command vehicle and the doctor dispatching victims to get information on their destination.

This System Suffers from Several Weaknesses

- The medical form may be damaged by multiple manipulations. Errors, omissions, trade, loss of documents can occur.
- Audio communication using walkie talkies imposes strict rules regarding transmission between users. The poor quality of audio transmission may induce errors in the transcription of medical assessments. These communications are often inaudible, wrongly used by stakeholders and eventually result in slowing down the procedure.
- The connection between the command vehicle and the hospital dispatch center is provided by the mobile network. In the even of a disaster this network may be saturated by the population and potentially neutralized by State order in cases of bomb attack (risk of triggering bomb by cell phone).

2. Materials: Suitable RFID Bracelets, Viewer

Plastic Bracelets including RFID Tag, RFID Reader, Tablet PC

- The RFID tags are easily available form of plastic strap. Many companies offer marketing labels incorporated in a carrier selected by the customer.

Passive tags 13.56 MHz R/W 2 to16K bits have been chosen for this use.

The memory chip can store any information, thus acting like a real medical board.

The frequency used allows a good compromise between speed of data transmission and distortion.

The detection of the chip will drive by about 2cm. This short distance also induced by the radio frequency of the label is a guarantee of safety.

At this distance, the reading can not be accidental, the labels are activated very selectively, thereby eliminating the risk of collision (activation of several labels at the same time and confusing information from the player).

- The market offers many robust tablets PC, watertight and suitable for extreme conditions in external environment. The touch screen allows one input with a stylus or a finger. Their loading battery lasts for several hours use. Some tablets PCs include an external RFID lector device.

Connectivity is possible via Wi-Fi, Ethernet, Bluetooth, GPRS, or 3G.

Software for Data Management

Many types of software are already developed by business leaders in the field of RFID, and most often are dedicated for industry use.

The chips memory contains a multiple tabs file. First tab will define the identification of the victim.

The second tab may be a page for balance rescuer.

The third tab will serve for the medical record, similar to the currently in-use paper card.

A final tab will match the final destination of the victim, i.e. hospitalization location, and details about transportation.

The computerization of data may allow the use of drop down menus for faster input information, and even automated quotations as for Glasgow Coma Scale scoring.

The development of software data management for medical evaluation may be "a la carte," according to needs of each department of care.

3. Local Network

On site of a disaster event involving a large number of victims, with deployment of a chain of medical emergency, the radio frequency technology for patient identification must be routed through a local network. Creating a local area network (LAN in English, short for local area network) supposes to connect multiple computers together, in order to achieve file transfers, sharing of resources, mobility and interactivity.

The wireless connection is the most interesting method because it allows deployment and quick operability independent of environmental conditions. The technology used for wireless LAN is based on IEEE 802.11.

The establishment of a wireless local network allows communication between multiple computers equipped with wireless adapter through a router. The IP (Internet Protocol) address of each network registered machine is pre-configured in IPv4. This minimizes the time to setup infrastructure on site.

The possible schema includes different tablet PCs connected to a central computer. This PC centralizes information of mobile phones, and represents the server database on the LAN. It receives and transmits files according to the requests of tablet PCs (clients). The data collected by different clients are stored on this server.

It can be equipped with a synthesis software which can collect data from different patient records, retrieve and record specific informations such as the number of identified victims, age distribution, types of emergency actions...

This server's local network may be physically located in the vehicle command post of the SAMU. The interchange format files is XML. (eXtensible Markup Language) , according to the standard "NC Software Security Civil adopted by the Working Group of INFOCERT GT399 (independent missionnée by A.F.NOR. Certification (French Agency for Standardization)). Standardization of software ensures the interoperability of the different actors of civil security. It also helps limit the file size so as not to saturate the bandwidth of transmissions.

The local server site may also provide information to another server in clone mode outside of the theater of events. Transmission technologies used are GSM / GPRS / EDGE, UMTS / 3G, and a particular digital radio electric method dedicated to national civil security (police, fire, SAMU). The latter is based on the I.N.P.T. (National Infrastructure Shareable Signals) whose network ANTARES (Adapted National Communications to the risks and Rescue) is shared with fire-fighters and paramedics. This network utilizes radio terminals based on the communication standard TE.T.RA. (Terrestrial Trunked Radio) industry group EADS (European Aeronautic Defense and Space). The advantage of this mode of transmission is that it relies on dedicated infrastructure and professionals as opposed to GSM or UMTS which could be saturated by default (public) or blocking (prefectural decision in case of suspicion of terrorist attack). Its main disadvantage is data rate limitation to 8kbits/seconde peak and an average of 2kbits/seconde.

Therefore the XML data files are designed by NC Software Security Preparedness to limit the size and avoid saturation of network bandwidth ANTARES.

4. RFID throughout the Medical Emergency Process

General Scheme

In this LAN configuration, each victim is provided with a plastic bracelet equiped with a radio frequency chip.

The first responder (first aid) uses his tablet PC with integrated reader chip and activates the chip bracelet. A folder appears on the PC. An identification tab allows to computerize data about identity of the victim. A 2nd tab, review first-aider can give quick information about the vital parameters. At the close of record, the RFID chip stores the medical record. The bracelet loads the entire file of the patient. This content is archived in the tablet PC and transmitted in real time to network. The process can be repeated at each step of the emergency chain.

The doctor reads the sorting bracelet with his tablet PC and the folder opens on the screen. He can easily categorize the degree of urgency. Transmission in the local network is identical, and the computer of the command vehicle therefore receives in real time the type and number of emergencies.

The process is similar at the level of advanced medical post. Doctors can reopen the file and meet diagnostic and therapeutic data. The set is transmitted to the LAN.

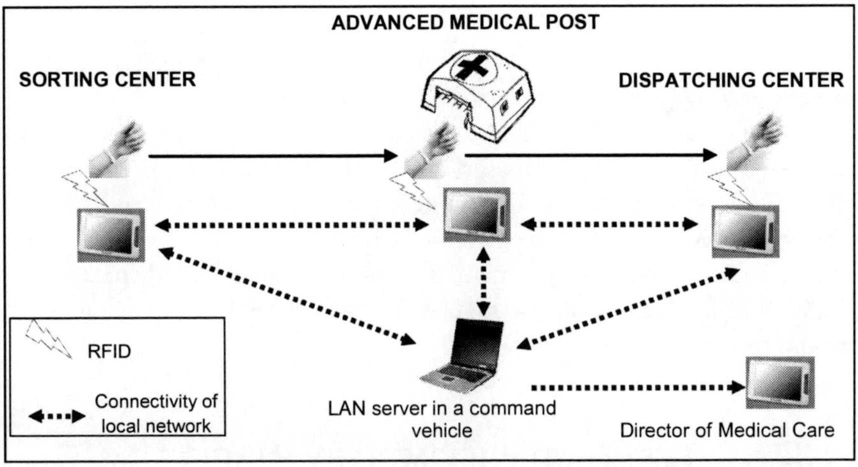

General Scheme.

Exchange of data from command vehicle to the draining center can transmit information about the final destination of the victim. Thus the physician in charge of the evacuation can quickly view the patient file (if the information has been recorded by the command vehicle) to assess destination and transport mode of the victim. This organization allows a fluidity in care with a medical record on a single RFID chip embedded. Different skateholders can read the chip and interfere on the record through a tablet PC- integrated RFID lector.

All mobile PCs would be held in local network. The local connectivity is still to be defined. A PC server database coordinates the network and represents a safety backup and synchronization system. Monitoring of patients is performed in real time at the command vehicle. When the destination of each victim is received by the VPC (from the medical center of regulation calls), this information is transmitted via the LAN to the evacuation center.

The computer in the command vehicle can access general synthesis software providing information on the number of people supported and their categorization (deceased urgencies, emergencies related). Such information can be sent by the LAN through tablet PC to the director of medical emergency, who sends real time information to the commander of relief operations and the warden of the state.

The director of medical care can track real-time activity of the chain of emergency, and get precise and detailed data to the government representatives.

Concept-with Offset Information

All data could technically be transmitted from the disaster-site local network, i.e. from the local server to another server in clone mode. Connectivity remains to be determined (see material and method-3) but exchange should be feasible even at great distance.

This outsourcing of information would provide the dispatch center hospital (the medical center of regulation calls) a large supervision of both sites and records.

Tests and Preliminary Results

1. Test on the Real Field Size

We observed the use of this technology to three times during exercise plan red.

Two of these simulated an explosion in a factory that caused dozens of victims. The third exercise simulated an accident in a tunnel with forty victims.

In 3 cases there was no risk NRBC. The first two exercises were performed daily in factories and the 3rd took place in a tunnel late at night.

The equipment included plastic bracelets with an RFID tag inside, and tablets PC with RFID reader. The local network was established as a GSM.

The software included a form for identification of the victim and a rescuer balance.

In parallel with this device the SAMU actors utilized also the medical paper chart for the identification of victims and establishment of medical records.

2. Problems

Users of this technology may encounter several problems, as followed:

- failure of GSM, which does not permit to establish a local network operations. In this case, the whole device was therefore completely inactivated.
- problem of duplicated sheets
- difficulty for users in handling the material (i.e. the entering procedure was too slow)
- inadequate lights on the screens making it difficult to read data and therefore unsuitable for outdoor conditions.

DISCUSSION

1. The Concept of RFID Computerization of the Chain of Emergency: The Expected Benefits

RFID is a reliable technology already widely used in industry. Its use in disaster medicine would then include all medical records on a simple plastic wristband, ensuring reliability and therefore security for transmission of such an information. Traceability is warranted because the system secures data at several levels (scalable backup on bracelet and computers are performed after each intervention). Thus the loss of information, omissions transcription, proofreading problems of handwritten information are minimized.

Adapting the system to a LAN achieves exchange of information in real time between different stakeholders, while removing the radio communications, or at least reducing them. Indeed radio communications for medical check-ups should be no longer useful. This allows:

- reserving radio–communications for other issues, such as requests for reinforcement in certain areas.
- improve the fluidity and speed of the overall intervention procedure.
- remove radio–transmission errors, and handwritten transcription.

Another key-point is that this technology might offer the opportunity to the director of medical care to retrieve reliable information in real time.

It could inform the prefect and the media about the amount and type of victims (age, gender ...), and their level of injury.

Relocation of informations (theoretical possibility) to the dispatch center via the digital radio network Antares results in:

- removing unreliable GSM connections in case of large-scale disaster. (saturated network, or even network failure from government source)
- giving the medical center of regulation calls a reliable overview of the on-site situation and real help for hospital needs adequation.
- Providing fast and reliable data to the physician in charge of victims dispatching, , through feedback within the same digital radio network

2. Areas for further Improvements

- The establishment of a local network must be fast with using an adapted and efficient connectivity. We must find the best compromise between speed and efficiency of transmission in external conditions. The system must be fast to implement, processing should be as simple as a quick start type on / of.
- The choice of material is crucial. We must choose materials appropriate to external conditions, which combine impact resistance, resistance to splash, usability and screen readability even if light exposure.
- The software of data management must be optimal. It is essential to develop a program as easy to play with and as fun as possible.

  Doctors should easily computerize information about diagnosis, to complete the medical card. It would be of particular interest to select a predetermined pattern area of the body and then access a menu displaying various pathologies or lesions to be selected. Some quotes (ex Glasgow GCS) and entered numbers on the vital parameters should also be available through specific lists and menus.

  Entering data about the treatment should follow the same principles; drop down menus should allow the selection of therapeutic large families, and then selection of the molecule and dosage.
- The extraction software synthetic data and offset information are still to be developed and tested.

3. Inevitable Disadvantages of this System

- Cost of Development: such a system implies high financial costs, due to the equipment itself, the various Tablet PC (minimum 4) and the network server. Further processing of such a project requires softwares with user licenses.
- This system needs staffs to be trained repeatidly and requires specific maintenance of the equipment. Because of its unfrequent use, financial investment in such a technology remains highly questionable.

4. Other Possible Application of RFID in Disaster Medicine

Inventory of Mobile Health Units of the SAMU

The mobile health posts are trailers for transport of drugs and careworkers to the scene of disaster medicine, in order to support a large number of victims. Two trailers park in our Department, for the care of 50 severely ill patients. The rest of the mobile medical station is stored in-hospital, at the central pharmacy and allows to support another 450 victims.

All materials and drugs require logistics and regular maintenance, scheduled inventories and much handling. It would be conceivable to use a RFID for automated inventory of these stocks.

An RFID tags can be placed on every box of drugs, each operative care. The chip can contain lots of information: name of medication, drug, dose, expiry date, exact amount in the box. This technique is currently wide spread in Industry, and allows a complete inventory of stock within minutes. This would highlight warnings on the expiry dates and also permit a quick after-use inventory.

Conclusion

The disasters are not uncommon. Populations are exposed to numerous and worsening risks, due to more intensive human infrastructures, transports, and increasing insecurity linked to acts of malevolence.

Our Department of Disaster Medicine focuses its efforts toward the improvement of medical practices in these emergency situations. It is our responsibility to develop strategies and technologies that might improve the rapidity of relief operations and ultimately save the greatest number of people involved in disasters.

The currently in-use paper medical chart and communication tools are of course still valid in case of red channel plan activation. Nevertheless, the computerization of the chain of aid and the use of RFID may represent an interesting substantial improvement of technics in the era of near future.

References

[1] Direction générale de la santé. Accidents collectifs, attentats, catastrophes naturelles: conduite à tenir pour les professionnels de santé. 2003 Novembre. Available from: http://www.sante.gouv.fr/htm/pointsur/attentat/guide.pdf

[2] Haut fonctionnaire de défense santé. Lutte contre le terrorisme nucléaire, radiologique, biologique et chimique: aspects sanitaires. 2004 décembre. Available from: http://www.sante.gouv.fr/htm/dossiers/nrbc/reponse/cadre_institutionnel/plans/risque_nrbc.htm

[3] Noto, R., Huguenard, P., Larcan, A. (1994). *Médecine de catastrophe.* (2ème édition). Masson.

[4] ALCOM Consulting et Newton. Vaureal Consulting pour le ministère de l'Economie, des Finances et de l'Industrie, Direction Générale des Entreprises. Etude sur les étiquettes électroniques et la traçabilité des objets. Panorama stratégique. 2007 Octobre. Available from: http://www.telecom.gouv.fr/fonds_documentaire/rapports/07/panorama_strategique.pdf

[5] Paret, D. (2008). *RFID en ultra et super haute fréquence UHF-SHF théorie et mise en œuvre.* Dunod.

[6] Hunt Daniel, V. (2007). *RFID: a guide to radio frequency identification.* Lavoisier.

[7] Paret, D. (2003). *Application en identification radiofrequence et cartes à puces sans contact.* Dunod.

[8] Briggs, L; Davis, R; Gutierrez, A; Kopetsky, M; Young, K; Veeramani, R. RFID in the blood supply chain--increasing productivity, quality and patient safety. *J Healthc Inf Manag.,* 2009 Fall;23(4):54-63.

[9] Lahtela, A; Saranto, K. RFID and medication care. *Stud Health Technol Inform.,* 2009;146:747-8.

[10] Jokela, J; Simons, T; Kuronen, P; Tammela, J; Jalasvirta, P; Nurmi, J; Harkke, V; Castrén, M. Implementing RFID technology in a novel triage system during a simulated mass casualty situation. *Int J Electron Healthc.,* 2008;4(1):105-18.

[11] Fry, EA; Lenert, LA. MASCAL: RFID tracking of patients, staff and equipment to enhance hospital response to mass casualty events. *AMIA Annu Symp Proc.,* 2005:261-5.

[12] Chan, TC ; Killeen, J ; Griswold, W ; Lenert, L. Information technology and emergency medical care during disasters. *Acad Emerg Med.,* 2004;11:1229-36.

[13] Noordergraaf, GJ; Bouman, JH ; van den Brink, EJ; van de Pompe, C; Savelkoul, TJ. Development of computer-assisted patient control for use in the hospital setting during mass casualty incidents. *Am J Emerg Med.,* 1996; 14(3):257-61.

[14] Bouman, JH ; Schouwerwou, RJ ; Van der Eijk, KJ ; van Leusden, AJ ; Savelkoul, TJ. Computerization of patient tracking and tracing during mass casualty incidents. *Eur J Emerg Med.,* 2000;7(3):211-6.

[15] Rafalski, E; Zun, L. Using GIS to monitor emergency room use in a large urban hospital in Chicago. *JMed Syst.,* 2004;28(3):311-9.

[16] Laurent, C; Beaucourt, L. Instant electronic patient data input during emergency response in major disaster setting: report on the use of a rugged wearable (handheld) device and the concept of information flow throughout the deployment of the disaster response upon hospital admission. *Stud Health Technol Inform.,* 2005;111:290-3.

In: Radio Frequency Identification ISBN 978-1-61122-416-0
Editor: Alison R. McAdams, pp. 93-106 © 2011 Nova Science Publishers, Inc.

Chapter 5

RFID Adoption in the Developed and Developing World

John Ayoade
School of Computing and Mathematical Sciences,
Auckland University of Technology, New Zealand

Abstract

Radio Frequency Identification (RFID) has gained international recognition based on many potential benefits it possesses. However, RFID adoption is facing a lot of challenges that many new and emerging technologies face. Some of the challenges peculiar to RFID adoption are cost, standardization, legality, privacy, time, fear and so on. This paper explores the challenges and benefits of adopting RFID in almost everyday life applications. This paper also discusses and gives recommendation on how to bridge the gap between the developed and the developing World in regards to RFID adoption.

1. Introduction

The most important organizational barrier to the adoption of new technology is cost. For new technology, both start-up costs and maintenance costs can be exorbitant.

In general, legal and regulatory barriers have focused on privacy and security of data transfer, Health Insurance Portability and Accountability Act (HIPAA), fraud and abuse, intellectual properties, and state licensing laws on adoption of new technology. Many of the problems in the legal and regulatory area concerning the adoption of new technology center on the lack of national and international standards and role of regulatory agencies. Time barriers center on acquiring, educating, implementing, using, and testing the efficacy and efficiency of new technology. Most organizations have little time to learn about new technology. Before most organization will accept new technology, they must be convinced that the new technology will not increase their workload and time in performing their task [11].

The benefits derived from the RFID adoption will be forces that can propel its full adoption. In order to establish this fact Table 1 is used to demonstrate the benefits of RFID over barcode system.

Table 1. Characteristics of RFID versus Barcode Technologies

	Barcode	RFID
Capacity	Have limited Space Are read-only	Hold substantial amounts of data Carry a unique identifier Allow for data reprogramming
Flexibility	Require line-of-sight scanning Allow for one simultaneous scan per read	Do not require line-of-sight scanning Read through most substances Support simultaneous reading
Accuracy	Require human intervention, which opens the possibilities for errors	Require little to no human intervention, which reduces errors
Durability	Can be easily damaged or destroyed	Withstand harsh environments (e.g. outdoors, chemicals, moisture, extreme temperatures)

Source: Deloitte, 2006.

2. Literature Review

2.1. The Importance of Bridging the Gap

The following are the reasons why the RFID adoption gap should be bridged between the developed and developing countries RFID adoption:

i. Health safety: traceability by the RFID technologies guarantee safety Many people in the developing countries suffer from the hand of fake drug manufacturers. Recently, news on adulterated baby food and medicine are very rampart. Most of these products are claimed to be produced in the developing countries however, they are not. They are locally produced. The RFID traceability of such products will guarantee safety.
ii. High demand of food and crops by the developed countries and accurate supply by the developing countries of such products Adoption of RFID system in the supply chain will help to meet the demand and supply of product from the developing to the developed countries and vice versa.
iii. Accurate and detailed information of such products RFID adoption will provide the accurate and detailed information of the product. This will not only give satisfaction but confidence in such products.
iv. Timely supply of goods to specific locations where they are needed It should be noted that various government regulations about food traceability do drive RFID adoption. For example, some African countries are building capacity for meat exports such as Botswana and Namibia [1]. The main components of Botswana's veterinary services that have enabled exports to developed countries are disease zo ning and traceability systems. Zoning is accomplished by means of a veterinary cordon fence and vaccinations, and the traceability system, in accordance with EU standards (EU food regulation no. 178/2002). The EU initiative came into effect in June 2005, with consumer safety as its first priority. The directive relates to all stages of food production, processing and distribution Directive forces any firm involved with food and beverages have a complete catalogue of anything that has been in contact with the final product, hence, ensuring its quality and safety. The law essentially aims to prevent fraudulent or deceptive practices in the food trade that result in

misleading information to the final consumer [8]. Botswana's animal and meat traceability system, the Livestock Identification and trace-back System (LITS), is similar to Namibia's. It utilizes a centralized electronic database to report and record cattle movements throughout Botswana. Rather than ear tags, however, LITS employs RFID-enabled boluses that are inserted into each cow's reticulum (second stomach) and contain scan-able information (owner's name, personal ID number, brand, brand position, sex, color, location, date of the bolus insertion). A radio frequency reader is able to detect the signal each bolus emits and transmit that information to central government and district computers [6].

2.2. Forces Propelling RFID Adoption

2.2.1. Mandates from Retailers

Radio frequency identification (RFID) is gaining wider adoption among manufacturers in the food industry, driven in large part by recent mandates from Wal-Mart Stores Inc., the largest food retailer in the United States, and Albertsons, one of the world's largest food and drug retailers. Wal-Mart and Albertsons announced that their largest 100 suppliers were required to begin using RFID technology by early 2005; and other major retailers have begun launching similar mandates.

Food manufacturers are looking for ways to begin implementing RFID solutions that meet the retailers' mandates with minimal initial investment, while still affording them the flexibility to expand processes as they learn more about the technology as it evolves some food manufacturers are taking a strategic approach to RFID, investing in the technology to take advantage of the increased visibility of tracking information in order to re-engineer their internal processes. Other food manufacturers offer RFID tagging, even when not mandated to do so, in order to competitively differentiate themselves and win new business from retailers that prefer products tagged with RFID. Because this technology is still new to the food industry, manufacturers that take the risk and begin working with RFID now will find themselves ahead of their competitors and better equipped to deal with the inevitable obstacles they will encounter once RFID compliance is required. But many food manufacturers continue to take a cautious approach, looking for a way to quickly meet mandates from retailers with minimum expense and minimum disruption to their manufacturing and distribution processes. They view RFID as a necessary cost of doing business – the "stakes" that they must put up to sit at the table and stay in the game with the major retailers. [9].

In the past years, RFID adoption was driven by mandates from the Department of Defense and from larger retailers such as Wal-Mart as expressed in the latter paragraph. However, Gartner states that the current market for RFID technologies is not only driven by compliance with such mandates, but the desire for business innovation. Companies are recognizing the value of RFID in business process innovation and view it as key to staying competitive. Globalization is another factor that has driven adoption of RFID in recent years, as businesses seek to decrease time to market products and services in new geographies [5].

2.2.2. Retail Supply Chain

The application of RFID in the retail supply chain has greatly contributed to the supply chain management and revolutionized the process by which products pass from manufacturer to retailer to consumer. The basic idea of RFID is a tiny computer chip placed on pallets, cases, or items. The data on chip can be read using a radio beam. RFID is a newer technology than barcodes, which are read using a laser beam. RFID is also more effective than barcodes at tracking moving objects in environments where barcode labels would be sub-optimal or could not be used as no direct line-of-sight is available, or where information needs to be automatically updated. RFID is based on wireless (radio) systems, which allows for non-contact reading of data about products, places, times, or transactions, thereby giving retailers and manufacturers alike timely and accurate data about the flow of product through their factories, warehouses, and stores [14]. The U.S. Department of Defense, with 43,000 suppliers is planning to overhaul its entire supply chain because it believes that RFID will reduce losses due to lack of information. The General Accounting Office substantiated the need in a December 2003 report that showed a $1.2 billion discrepancy between the material shipped and the material received in Iraq by the Army [17]

2.2.3. Food Traceability

Regulations that mandate the ability to trace the origins of food are being put in place to control the food chain, solve food scares quickly, and prevent bioterrorism. According to a new brief by Forrester Research, tagging food with RFID tags will help manufacturers and retailers comply.

To address the problem of food traceability, retailers and consumer packaged goods firms should use RFID tags to: meet traceability compliance deadlines; integrate agricultural firms into the food chain; slash product recall costs with case-level RFID tags; and probe RFID tags benefits with a clear business case. Regulations don't specify the use of RFID tags to comply with food safety regulations, but using them to find goods in distribution centers, retail stores, and trucks in transit will help firms respond within the predefined time limits to any official inquiry. Traditional tracing technologies, such as ear tags in animals and barcodes on packaged meat, provide data for one or two steps in the food business but don't connect up the entire food supply chain. To achieve multistep, forward-and-backward food traceability, it is advised that firms should turn to vendors that offer RFID tags to link unprocessed agricultural products to retail-environment-ready consumer products. Firms can use RFID in multiple strategies to get a better understanding of: how goods flow through the supply chain; stock control; anti-theft strategies; buying pattern analysis; and food tracking [10]

Outbreaks of food borne diseases have focused attention on food safety and traceability of food products/Public health, food and supplies in United States and European Union have issued guidelines requiring manufacturers to enhance food safety by strict quality control and better traceability of food products in the supply chain. This includes deploying identification and tracking technology for enhanced visibility and traceability. RFID is the most promising technology for meeting these requirements as it can potentially enable tracking of food products from farms to the platter. With RFID it's possible to track the chain from agrochemicals manufacturing, through the farms, food processing units, distribution and retail chains. This makes it possible to effectively trace, identify and recall consignments suspected of contamination or rendered unfit for consumption. It can also serve as a check against intentional contamination. RFID can help companies minimize the negative impact on their businesses in the event of contaminations and also outbreaks of food borne diseases like the recent bird flu and mad cow disease outbreaks. [16].

2.2.4. Authenticity

Some other areas where the RFID could be of benefits to the developing World are the enforcements of the production of the authentic drugs and household goods for both consumption (food) and non-consumption goods (such as car-spare parts, clothes, household equipments). Some are manufactured in the developing countries and claim to come from the developed countries. With the deployments of the RFID system such atrocities will be deterred. Actually, there have been occasions where adulterated drugs and processed can baby food had kill many in the developing World.

According to [7], many marketplace including large wholesale distribution customers, believe that the benefits of RFID can be significant. Most of these large customers are taking highly aggressive –or offensive approaches to capture RFID benefits by investing in infrastructure and mandating supplier participation. Others in the marketplace who are skeptical of RFID-enabled benefits are taking a more cautious –or defensive approach.

2.2.5. ICT Giants Contributions

Few ICT giants such as Nokia Simens Network, Microsoft and Cisco have contributed over 125,000 US dollars to establish an International Telecommunications Union (ITU) fund to bridge the standardization gap between developed and developing countries. ITU recently announced the fund that would be used to support forums, tutorials and workshops, participation of delegates from least developed countries in meetings, the hosting of meetings in developing countries, surveys and study programs. Bridging the standardization gap means allowing easier participation in the standards development process which in turn allows developing countries to profit from access to new technology development and ensures that their needs are taken into account in the development of standards [12]. Increasing developing country participation in the standards making process is not a short term fix. It will give developing countries a voice in the development of next generation ICTs and sow the seeds of a truly equitable information society."

3. Discussion

3.1. Challenges to the RFID Adoption

3.1.1. Telecommunication Infrastructure

This factor will be one of the major challenges because by using RFID in retail scenario, goods will be located along the entire process chain – from production all the way through to the shelf in the store. Managing orders can be optimized, losses reduced and out-of-stock situations avoided, assuring an even more consistent availability of goods for the customer. Similarly, in future, it is predicted with the help of ubiquitous technology a user could be informed automatically about the status of the food in his/her refrigerator (smart appliances). The Smart home would maintain data on inventory levels as well as consumption. Periodically, the consumer would give permission to his/her home server to upload her new shopping list to the system. In order to realize this, the present condition of the telecommunication infrastructure in the developing countries might be a great impediment to the actualization of the RFID adoption [13].

3.1.2. Privacy

There are many interesting RFID applications but consumers privacy violations had been a focal point. For example, there could be a number of applications within education environment related to security, privacy and the identification of staff and students, but this, of course, is controversial as there are serious concerns over privacy. RFID technology can be used for identification and location tracking of a person carrying the tag (which can be embedded into an identification card) and can be used to verify a person's right to enter a particular building or even to access a service. For example, in the US, RFID-enable cards are used to automatically register students at lectures, and in China, nearly 3,000 universities have installed RFID tag readers, for use with RFID-enabled student identification cards, contactless library applications, and to reduce train travel fraud [4]. In a nutshell, both the benefits of such RFID application and challenges should be weighed and a balance should be maintained.

3.1.3. Standard

Standard is another important impediments that are needed to be look into. A great many RFID applications have been proposed and some have been deployed although, some are still on the pilot stage. In order to have a complete adoption of RFID system a lot of research works are still needed to be put in place. Although, some institutions and organizations are working on the standardization and interoperability of the RFID systems and protocols but this might take a longer time to be able to come out with a suitable one most especially when there is no tangible plan to bridge the gap between the developed and the developing countries efforts.

3.1.4. Health Issue

The RFID equipments' (readers and tags) manual advises users to be careful about the RF (radio frequency) radiation that the reader produces. Although, it is rare to find publications confirming incidents of havoc that the radiation from the RFID readers have caused. However, few articles have been published online that expressed their concern about some users that actually stated that they have symptoms of some health issues like headache after they have used some equipments that produces radio frequency radiation. According to American Cancer Society, non-ionizing radiation is low-frequency radiation that does not have enough energy to cause ionization in tissues, but may cause adverse health consequences in other ways. There are a number of non-ionizing radiations such as, ultraviolet rays, visible light, electromagnetic fields, infrared radiation, microwaves, and radiofrequency radiation (radio waves). Of all the types of non-ionizing radiation, only ultraviolet rays have been established as a cancer-causing agent [2].

3.1.5. Data Handling, Distribution, and Mining

There are many issues on how the data in RFID is handled and distributed. This depends on the kind of RFID application is planned to be deployed. Moreover, some RFID applications are being adopted into other applications without modifying the architecture of the initial applications. In this kind of situation there could be privacy or security loop holes in the adopted application. Such loop holes could also lead to mishandling or intrusion into the data being distributed for example along the supply chain line.

Schmidt et al stated that comprehensive models that support handling of data mining as well as correlation over space and time will improve the usefulness of data collected. Of particular interest where to store data (on the tag, in the supplier's systems, in customer systems, or at a trusted third party) and how to organize sharing along the supply chain [15].

3.1.6. Benchmarking and Usage Contexts

Schmidt et al believes several projects showed that individual tags, even from the same manufacturer, behave differently. So, it would be very useful to have benchmarking tests that are accepted and that we can use to evaluate how different tags perform with different readers. Such tests would need to take into account the materials onto which tags are attached and reading environments. We need to understand what determines tags' suitability in different contexts and what the basis is for different quality standards [15].

3.1.7. Security and Economic

Schmidt et al [15] also expresses that RFID involves many security aspects, from encrypted storage on the tag to secure radio protocols or password protection of tags. For many application scenarios, we need to provide practical and secure solutions before we can expect RFID use to increase.

For many companies, the cost of tags and supporting RFID equipment (e.g. readers, encoders, and IT hardware) remains a barrier to adoption. At present, tags are far more expensive than barcodes. However, growing demand is expected to bring costs down from the current 30-80 cents per passive tag to an estimated 5-15 cents per tag over the next four years.

4. RECOMMENDATION

Almost all the benefits that RFID could provide in the developed World could also be provided in the developing World. However, adoption of RFID in the developing countries will face more challenges that their developed World counterpart faces.

The following are the challenges and recommendations:

4.1. Cost

Cost of adopting RFID in the developing countries is a great challenge because of the cost of tags and readers to replace barcode systems which are presently used in some of the developed and developing countries. Only few organizations use barcode system in the developing countries though. Some organizations believe perhaps its better to start with the barcode system before moving to the level of adopting RFID system just because of cost implementation. However, the reward of adopting RFID system supersedes the cost of deploying it. RFID pilots had been deployed in most developed countries and as people and industries see the potential that RFID deployment has they are more tends towards seeing the benefits and rewards of adopting the RFID than the initial cost of deploying it.

The adoption of cellular phone is a living proof to this. In retrospect, as of 1990 nobody believes that the cellular phone will be widely accepted in the developing countries as it is presently. The reason being that people in the developing countries forseen the benefits and importance of cellular phone which was difficult to believe that cellular phone could be so accepted and adopted because of the cost it attracted for deploying and adopting it then.

However, today to utmost surprise the cost becomes secondary and the benefits are seeing as the primary issues.

4.2. Standardization, Awareness and Training

There have been some moves in the developed countries on how to standardized the deployment of RFID. The reason being that RFID is a global technology just like the way the Internet has become global technology and it has no boundary. In the nearest future RFID would be the same. On the contrary, developing countries should be part of the standardization policies and protocol.

Awareness and training on the benefits of RFID is being part of what have made RFID adoption well accepted in the developed World unlike in the developing World where only few people are aware and are opportune to enjoy RFID training.

5. CONCLUSION

Most developing countries retailers and consumers of goods are not aware of the potentials of RFID and those that do have the awareness are faced with the challenges that have been discussed above.

One of the reasons why the gap should be bridged between developed and developing countries is because many world's largest retailers are deploying RFID systems not only to reduce their labor and inventory costs but also they think that their revenues will increase by limiting the out-of stock items throughout their chain stores around the World. Therefore, in order for most world largest retailers fulfill these goals bridging of the necessary gap is necessary. Most developed countries fall in the temperate climate zone where cropping season is limited to the summer months. On the other hand, crops can be grown throughout the year in Africa as the continent is endowed with a tropical climate and high altitude regions, increase in consumer demand in developed countries for out of the season fresh fruits and vegetables has opened a niche for African countries to produce these crops for export during the void period at attractive prices. As a result procurement of fresh fruits and vegetables by the developed countries from the developing countries is a must. However, increasing demands for food safety and traceability of products supplied by the developing countries to the developed countries have faced increased scrutiny as to their origin, date and other necessary information of processing and dispatch [3]. Actually this could be handled if the RFID technologies could be adopted. Moreover, there have been concerns of food and crop exportation from a country to the other. Not so long ago, a case of mad cow disease hit the U.S.A and this led to the importation ban on beef from the U.S.A by the Japanese government. However, the ban was lifted due to the tracking and traceability evidence from the U.S.A as a result of RFID technologies.

5.1. The Strategies to Bridge the Gap

The following are some guidelines and steps the developing countries need to take to bring their RFID adoption levels up to the level of developed world:

i. Formulate and establish a body to bridge and integrate RFID adoption between international RFID technology development and local technology adoption. One of the ways the developing countries can bring the RFID adoption to the level in which the RFID adoption is in the developed countries is to formulate and establish a technical committee and corporate organizations that have interest in RFID adoption. The role of such a committe is to study RFID adoption in the developed countries and see how such strategies used in the developed could be incorporated in the developing countries so that the huge gap could be bridged. This body will actively participate in the RFID International standardization activities.
ii. Collaborate with both developed and developing government and industry for technology. Presently in the developed countries there are different RFID standards. For example U.S. and Europe has a standard, Japanese has its own standard and even China has started to have their standard. This means that developing countries need to develop their standards and the reason being that for the RFID to be fully adopted in any country standardization is very important. Apart from developing countries developing their RFID frequency standards. There has to be collaboration efforts from the developing countries with the developed countries. This technical committee from the developing countries will bring and implement RFID technology

standards based on their findings from their study from the developed countries.

iii. Develop technology adoption life cycle (part of this will be training of industry interested, identify industries that will need to adopt it on time). The development of adoption life cycle would be a roadmap for developing countries to get to the stage where the developed world RFID adoption is because it will educate the interested individuals, researchers, organization and governments in how to adopt RFID in their respective fields.

iv. Identify the reason, advantages, disadvantages which should be justified with tangible reasons. The formulated committee will be able to identify the potential advantages and disadvantages of adopting RFID in developing World.

REFERENCES

[1] Cabrera R., Cochran M.,Dangelmayr L., D'Aguilar G., Gawande K., Lee J., Speir I., Weigand C. *"African Capacity Building for Meat Exports: Lessons from the Namibian and Botswanan Beef Industries"* http://bush.tamu.edu/research/workingpapers/kgawande/AfricaBeef.pdf [Africap 2008]

[2] *"Radiation Exposure and Cancer"* http://www.cancer.org/docroot/PED/content/PED_1_3X_Radiation_Exposure_and_Cancer.asp [Cancer Society 2006]

[3] Amondi R. *"Bridging the Traceability Gap in Food Exports from Africa to Europe"* EU RFID Forum 2007 [Amondi 2007] http://www.rfidconvocation.eu/Papers%20presented/Business/EXPLORING%20RFID%20FROM%20THE%20AFRICAN%20FARM%20TO%20EU%20FORK.pdf

[4] Collins *"Smart labels for Higher Education" RFID* Journal http://www.rfidjournal.com/articlereview/666/1/1 [Collins 2003]

[5] *"RFID Adoption Trends in the IT Channel"* May 2008 [CompTIA 2008] http://www.comptia.org/sections/research/white%20papers/SecurityWP4-08.pdf

[6] *Department of Animal Health and Production (DAHP)* 2005; European Commission 2000 [DAHP 2005]

[7] *"Understanding RFID Challenges and Risks"* 2006 [Deloitte 2006]

[8] *"RFID Adoption and Implications"* http://www.ebusiness-watch.org/studies/special_topics/2007/documents/Study_07-2008_RFID.pdf [Ebusiness 2007]

[9] *"Factors to Consider When Choosing a RFID System"* [Food Manufacturing 2007

[10] *"Meat Traceability Food RFID: Tracing the origin of food supply, including meat,* with RFID" http://www.yenra.com/meat-traceability-food-rfid [Forester 2004]

[11] *Identifying Barriers to the Adoption of New Technology in Rural Hospitals:* A Case Report http://www.pubmedcentral.nih.gov/articlerender.fcgi?artid=2047308 [Garrett 2006]

[12] "*ICT Giants Contribute to bridging the Standardization Gap"* 2008 http://www.itu.int/ITU-T/newslog/ICT+Giants+Contribute+To+Bridging+The+Standardization+Gap.aspx [ICT 2008]

[13] Kumar R. , Chartterjee R. *"Shaping Ubiquity for the Developing World"* [Kumar 2005]

[14] Loebbecke C. *"RFID Technology and Applications in the Retail Supply Chain:* The Early Metro Group Pilot" http://www.mm.uni-koeln.de/team-loebbecke-publications-proceedings/Conf-081-2005-RFID%20Technology%20and%20Applications%20in%20the%20Retail%20Supply%20Chain.pdf[Loebbecke 2005]

[15] Schmidt A. Spiekermann S. Gershman A. Michahelles F. "*Real-World Challenges of Pervasive Computing"*http://csdl2.computer.org/comp/mags/pc/2006/03/b3091.pdf [Schmidt 2006]

[16] "*Enabling Traceability"* http://www.tcs.com/SiteCollectionDocuments/White%20Papers/Enabling%20Traceability.pdf [TCS 2004]

[17] Tegtmeirer L. "*RFID Knowledge Enabled Logistics: Supply Chain Management"* http://www.itu.int/osg/spu/ni/ubiquitous/ [Tegtmeirer 2004]

In: Radio Frequency Identification ISBN 978-1-61122-416-0
Editor:Alison R. McAdams, pp. 107-123

Chapter 6

ROADMAP TO RADIO FREQUENCY IDENTIFICATION AND LIBRARY IMPLEMENTATION: A REVIEW*

***Viplove Goel, Anil Chauhan, Shailesh Goswami, Rohit Bankoti and B.K. Kaushik Govind*†**
Ballabh Pant Engineering College, Pauri Garhwal, Uttarakhand, India

ABSTRACT

In this paper we have overviewed the latest technology RFID (An automatic identification procedure) by comparing it with the ongoing technologies like barcode, smart cards on various parameters. There is a brief chronological history about the development of this technology starting from 1800 to till date developments. We have also discussed about the essential features of the RFID system, which are tags, readers, application software. We then have illustrated in detail various types of tags:-Active, Passive, then about readers and how the tag and reader are coupled. These mainly work at 13.56MHz frequency, under international

---

* A version of this chapter was also published in *Journal of Current Issues in Media and Telecommunications,* Volume 2, Number 1, published by Nova Science Publishers, Inc. It was submitted for appropriate modifications in an effort to encourage wider dissemination of research.

† Email: brajesh_k_k@yahoo.com

standards. The most important advantage of RFID is its data security which is also being dealt with along with its application, advantages and disadvantages. This technology is widely being used in different fields like health, animal identification, passports, making it a ubiquitous identity card for every fields, libraries etc.

1. Introduction

In recent times automatic identification procedures (Auto ID) are gaining popularity in fields containing information about person, animal, goods and products in transit. The omnipresent barcodes which introduced the concept of identification system are fast being replaced by new technology known as RFID. It is an acronym for Radio Frequency Identification. These provide the facility of contactless identification and increased the security by using features like data security and cryptography. The various automatic identifications are listed below with their some efficient features.

Barcode System

The barcode is a binary code comprising a field of bars and gaps arranged in a parallel configuration.[1] The sequence is made up of wide and narrow bars and gaps, which can be interpreted numerically and alphanumerically. They have the added advantage that if reader does not work they can be read by naked eyes.

Biometric Procedures

Biometry is the general term for all procedures that identify people by comparing unmistakable and individual physical characteristics.[1] The various technologies are Voice Recognition and Fingerprinting Procedures (dactyloscopy).

Smart Cards

These are electronic data storage system with microprocessor card incorporated into a plastic card of size of credit card for convenience. Data transfer between the reader and card takes place bidirectionally. They are vulnerable to corrosion, wear and tear and frequently expensive to maintain. Some of the cards use EEPROM and some use Microprocessor cards.

2. History of RFID

Before going deeply into RFID let us be retrospective and look at the history and know about the development into this recent technology. The 1 800s marked the beginning of the fundamental understanding of electromagnetic energy. The works of Maxwell were confirmed by German scientist Heinrich Rudolf Hertz in 1887, and is credited for the first person to transmit and receive electromagnetic wave. After that the development of radar to detect radio waves helped the scientist in 1970 to develop intended applications for animal tracking, vehicle tracking. 1980's became the decade for full development of RFID technology for short range systems for personnel works. In 1990's the progress didn't slow down since new technologies expanded the horizon for RFID. For the first time the Schottky diodes were fabricated on a regular CMOS integrated circuit. The construction helped in the construction of microwave RFID tags that contained only single integrated circuit., this increased the pursuit for companies like IBM and Single Chip System. There is emergence of various standards for RFID deployment in various fields like Library, Health Care, Toll Bridge and everyday part of life.

3. Essential Parts and Features of RFID System

A basic RFID application system has several components, including RFID tag, read/write terminal (RF sensor), RF controller, and managing software in the PC[2].When the reader and tag are in same frequency then it will come into work area RFID system can work in full duplex, half duplex and sequential modes. The block diagram of general RFID system is shown in Figure 1.

The above figure outlines the block structure of the system. It is evident that the system is centered around the reader IC (MF RC500) and the microcontroller (89C668). A read/write command initiated by the user is supplied to the reader IC via the microcontroller. The reader IC provides the necessary signals to generate a RF field using the tuned antenna-capacitor circuit. When the passive tag is brought within the field it is energized and authenticated via code. It derives just enough power to retransmit the requested information. The data received is in the modulated and encoded format. The reader IC is responsible for demodulation and decoding of data. This received stream of data is stored temporarily in the microcontroller memory and sent out via the serial port for display as required at the user end. The various integral parts like tags and readers have been discussed later in the paper.

The data capacity ranges from few bytes to several kilobytes. The procedure to store data is to store in inductively coupled EEPROMS but they have the disadvantage of high power consumption. SRAM's are also being used in storing the data. Different transmission frequencies are classified into the three basic ranges, LF (low frequency, 30–300 kHz), HF (high frequency)/RF radio frequency (3–30MHz) and UHF (ultra high frequency, 300MHz–3 GHz)/microwave (>3 GHz). A further subdivision of RFID systems according to range allows us to differentiate between close-coupling (0–1 cm), remote-coupling (0–1 m), and long-range (>1 m) systems. The different procedures for sending data from the transponder back to the reader can be classified into three groups: (i) the use of reflection or backscatter (the frequency of the reflected wave corresponds with the transmission frequency of the reader → frequency ratio 1:1) or (ii) load modulation (the reader's field is influenced by the transponder → frequency ratio 1:1), and (iii) the use of subharmonics (1/*n* fold) and the generation of harmonic waves (*n*-fold) in the transponder[1].

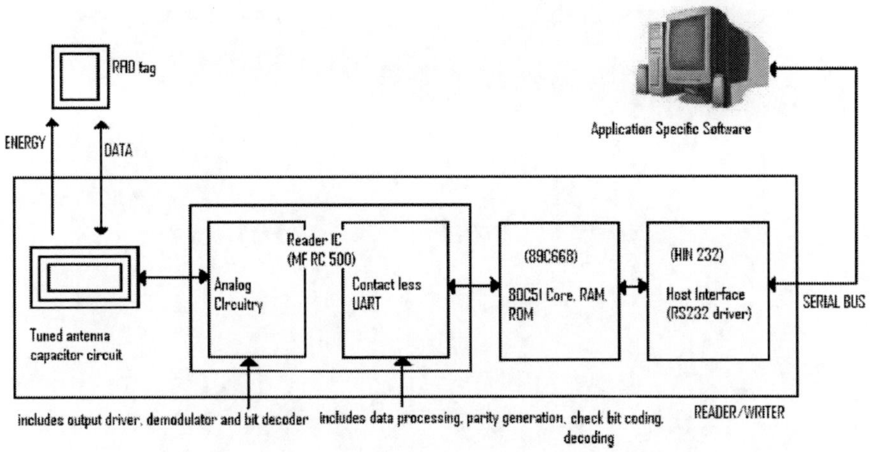

Figure 1. Block Diagram of RFID system [3].

4. RFID Tags

The main components of the RFID are transponders (tags) and readers. RFID tags have an antenna and silicon chip. Antenna is used as a radio receiver and radio modulator that send back the response signal to the reader. Silicon chip is used as a memory (stores data). In very simple systems, the transponder's data record, usually a simple (serial) number, is incorporated when the chip is manufactured and cannot be altered thereafter. In writable transponders, on the other hand, the reader can write data to the transponder. One broad classification of RFID tags is whether they contain a microchip or not. "Chip" tags contain an integrated circuit chip, whereas "chipless" tags do not. Chipless tags are less expensive to make and may store up to 24 bits of information – which provides enough memory for a company's internal use, such as on a shop floor or within a warehouse. However, that is not enough for mass-market applications [5].RFID tags are also classified as passive tags and active tags. Passive tags have no own power supply (battery) and are powered by the radio signal. A RFID reader sends radio signals that wake up the tag within a range, enable it to transmit the information that is stored on it. "Active" tags have their own battery which reduces their life cycle. They either broadcast their information without being interrogated by the reader, or stay quiet until triggered by a reader.

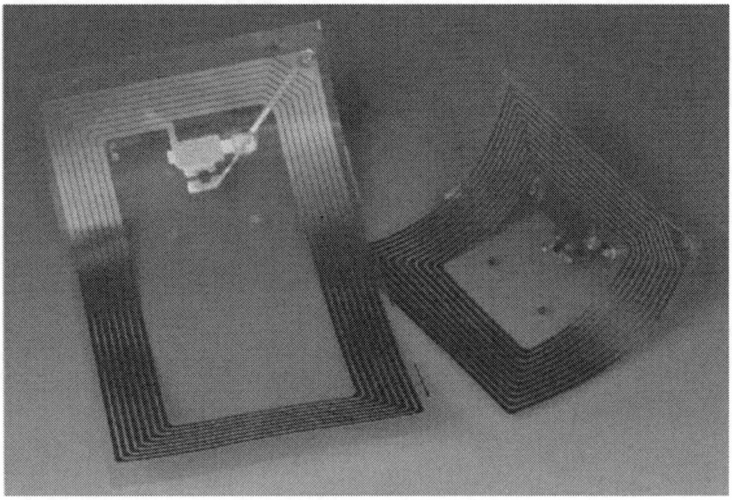

Figure 2. RFID tag.

4.1. Active Tags

Active RFID tags, which have both an on-tag power source and an active transmitter, offer superior performance. Because they are connected to their own battery, they can be read at a much higher range – from several kilometers away. But they are larger and more expensive. Active RFID tags are suitable for manufacturing, such as tracking components on an assembly line, or for logistics –primarily where the tag device will be reused [5].

4.2. Passive Tags

Passive tags are simplest, smallest and cheapest among the all kind of tags. Passive tags have no power source and no on-tag transmitter, which gives them a range of less than 10-metres and makes them sensitive to regulatory and environmental constraints [5]. They also have a very long life span unless they are damaged or torn therefore are widely used in many different retail items. Passive tags can operate at low, high, ultrahigh, or microwave frequency but are ore exposed to electromagnetic noise.

Passive tags are triggered with the help of the reader's transmitting power. The reader first interrogates the tag with a query through electromagnetic waves which when inductively coupled with the antenna of tag energizes it. As shown in Figure 3 the transponder antenna gets its energy from the field generated by reader. This energy enables the transponder to communicate with the reader or computer. Due to the penetration of the emitting field of reader a small voltage is generated in the transponder coil. This voltage is rectified and serves as the power supply for the data-carrying device (microchip).

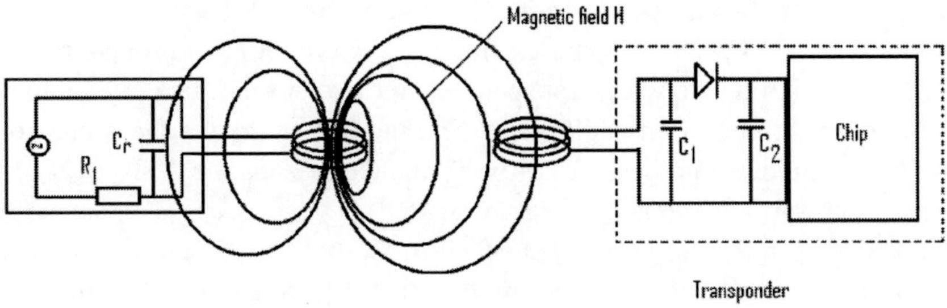

Figure 3. Power supply to an inductively coupled transponder from the energy of the magnetic alternating field generated by the reader.

A capacitor C_r is connected in parallel with the reader's antenna coil, the capacitance of this capacitor being selected such that it works with the coil inductance of the antenna coil to form a parallel resonant circuit with a resonant frequency that corresponds with the transmission frequency of the reader.

5. Readers

This means that all reader and transponder activities are initiated by the application software. In a hierarchical system structure the application software represents the master, while the reader, as the slave, is only activated when write/read commands are received from the application software. The reader's main functions are therefore to activate the data carrier (transponder), structure the communication sequence with the transponder, and transfer data.

Readers in all systems can be reduced to two fundamental functional blocks: the control system and the HF interface, consisting of a transmitter and receiver. As shown in Figure 5. HF interface, which is shielded against undesired spurious emissions by a tinplate housing is located on the right hand side. HF interface generate high frequency transmission power to trigger transponder, modulate transmission signal to send data to transponder and receive and demodulate signal sent by transponder. The HF interface contains two separate signal paths to correspond with the two directions of data flow from and to the transponder.

The control system is located on the left-hand side of the reader and, in this case, it comprises an ASIC module and microcontroller. Control unit communicates with microprocessor and execute the command. It also communicate with the transponder. The control unit is usually based upon a microprocessor to perform complex functions. Cryptological procedures, such as stream ciphering between transponder and reader, and also signal coding, are often performed in an additional ASIC module to relieve the processor of calculation intensive processes. For performance reasons the ASIC is accessed via the microprocessor bus (register orientated) [1].

In order that it can be integrated into a software application, this reader has an RS232 interface to perform the data exchange between the reader (slave) and the external application software (master).

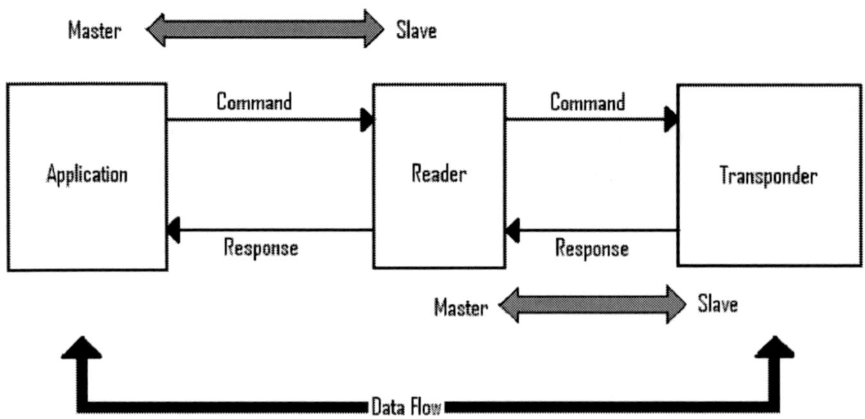

Figure 4. Master–slave principle between application software (application), reader and transponder.

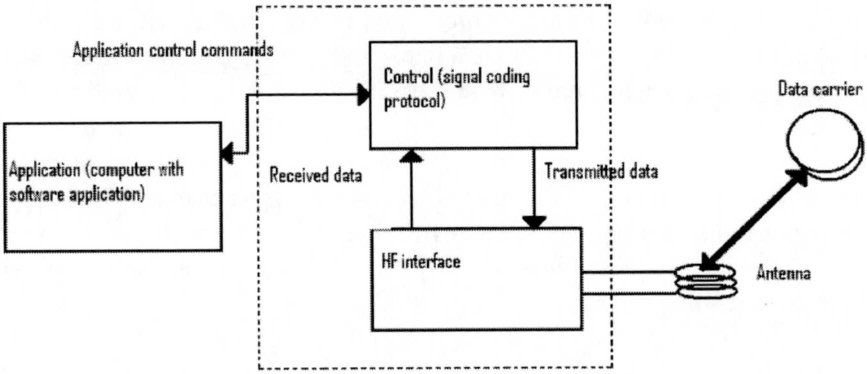

Figure 5. Block diagram of a reader consisting of control system and HF interface.

6. Data Security

RFID systems are being used in high security applications such as for making payments for or issuing tickets. High security data must have protection against these attacks:

- Unauthorized reading of a data carrier in order to duplicate and/or modify data. [1]
- The placing of a foreign data carrier within the interrogation zone of a reader with the intention of gaining unauthorized access to a building or receiving services without payment. [1]
- Eavesdropping into radio communications and replaying the data, in order to imitate a genuine data carrier ('replay and fraud'). [1]

When selecting RFID system, consideration should be given to cryptological statements. There are generally three methods for encryption (i) Mutual authentication procedure (ii) Authentication using derived keys (iii) Encrypted Data Transfer

7. Advantages and Disadvantages of RFID

The first and the foremost advantage of RFID is that there is no need of line of sight for scanning the tags. We also simultaneously read many tags with the help of anti collision theorem, and of course the cost and work of labor is reduced. The information is reliable and secured and the whole system is robust and durable. There is higher asset utilization as compared to other automatic identification systems with improved inventory management

The various disadvantages in implementation of RFID is its high cost compared to the existing Bar Codes. There is lack of trained staff present for this application, there are interference limitations, there is also problems in deployment in different supply chain partners. There is also some concern about the consumer privacy.

8. Applications of RFID

Some libraries have implemented RFID systems to facilitate book checkout and inventory control and to reduce repetitive stress injuries in librarians [3]. By exploiting RFID tags in garments and packages of food, home appliances could operate in much more sophisticated ways. Washing machines might automatically choose an appropriate wash cycle. Personal medical RFID card with prescription and RFID tagging of pharmaceuticals. In retail shops, consumers could check out by rolling shopping carts past point-of-sale terminals [3].

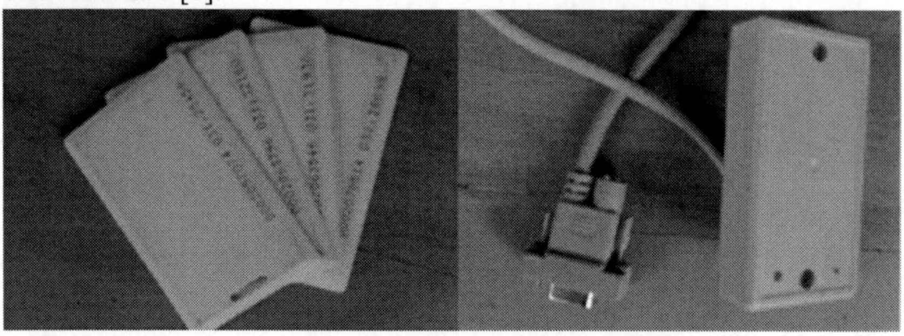

Figure 6. Passive tags and GP-220 Reader used in implementation.

9. RFID Library Implementation – A Case Study

9.1. Introduction

Radio Frequency Identification Technology is replacing the traditional barcodes on library items (books, CDs, DVDs, etc.). The tags present on the item are 50mm × 50 mm and also acts security device, taking the place of more traditional electromagnetic security strip. The various needs of library are proper control, easy identification, prevention of loss/theft, easy tracking, speedy issue/return, reduction of human work [3]. The basic three things needed are tags, reader and application software. These things provide the basic setup of library system, however to avail the advantages we need to have anti theft counter so that we can protect our library from miscreants. In this case study we will just explain the simple working of library with the above three basic parts.

9.2. Methodology

In the old omnipresent technology of barcodes we have seen that the person selects the book from the rack and comes to the librarian, who brings the reader near to the barcode and from there the information about the book is read and then the information from the student ID card is also added from the id card. This is a long tedious task reducing the speed of issuing and returning of the book. When the book is returned its position has to be found out, and then only we can place it there and there may be many human errors. Introduction of RFID into library system we can surpass this orthodox time consuming technique. A passive tag is positioned on the book which contains the information about the book like, book number, book name and author. In addition to this an ID card with passive tag contains the information about the student like, name, year of studying, branch and other information. These can be collectively put on the RFID reader which gathers the information about the book and the student and these are transferred to the account of the student automatically with one click of mouse. When the book is returned the procedure can be applied and the position of the book to be kept in the rack can be automatically detected through a RFID reader specially designed for this work. In earlier systems we have to deploy a person to stop the theft of books but if we install a anti theft gate then if the book is not being issued it will start an alarm reporting about the theft of the book.

9.3. RS-232 Port

The serial port used in computer is full duplex device which allows two way communication but in our implementation we are using it in half duplex mode with pins 1, 3, 5. It can send or receive data simultaneously. Serials ports on most computers use RS-232 as standard .In a SCI (Serial communication interface), serial communication between processor and other devices takes place in a full duplex, asynchronous, NRZ (Not Return to Zero) manner. Asynchronous means no synchronization and does not require sending and receiving idle characters as in case of synchronous communication. Synchronous communication allows faster data rate than asynchronous because there is no need of additional bits to represent start and end of transmission of data byte. It has one advantage over synchronous communication that processor does not deal with additional idle characters.

Free line is defined status of logical one. Start bit represents start of transmission and having status of logical zero. Data bits follows start bit and after bits stop bit of logic one status comes. Stop bit represents end of transmission and having status of logical one.

9.3.1. Functions of Signals in RS-232[6]

1) CD (Carrier Detect): Signal at pin generated by DCE (Data Communication Equipment) that connection is established with remote equipment.
2) RXD (Receive Data): Data transmitted from DCE to DTE (Data Transmitted Equipment)
3) TXD (Transmit Data): Data transmitted from DTE to DCE.
4) DTR (Data Terminal Ready): This pin indicates to data set that DTE (Data Terminal Equipment) is on.
5) GND (Ground): This pin is reference point for all interface voltages.
6) DSR (Data Set Ready): This pin indicates that DCE is on.
7) RTS (Request To Send): Signal received at this pin is from transmitting computer (DTE).
6) CTS (Clear To Send): DCE (Data Communication Equipment) is ready to transmit

9) RI (Ring Indicator): This signal is sent from modem to the computer. Its purpose is to indicate to the computer that phone line is ringing.

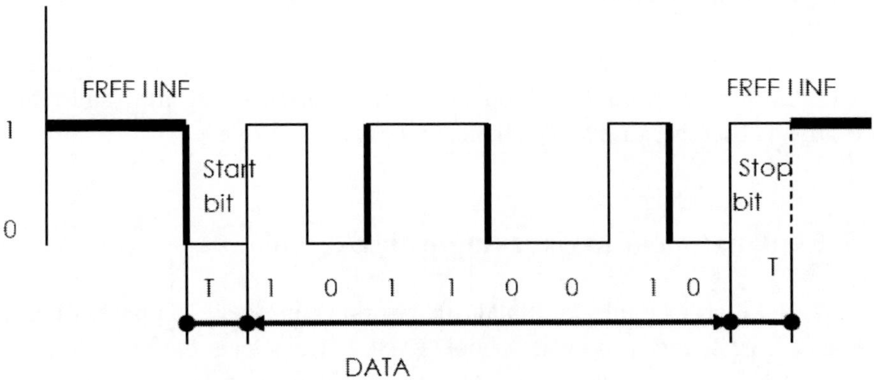

Figure 7. NRZ Asynchronous Data Format.

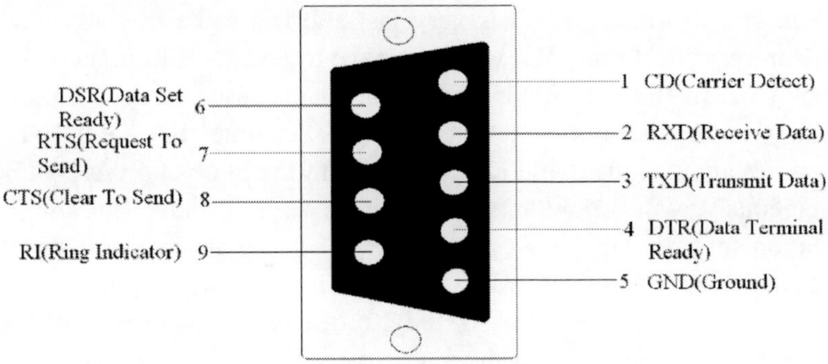

Figure 8. Pin diagram of RS-232 serial port.

9.4. System Overview

To show the above we have taken four tags numbered “0002063546 031, 31930”, “0002063796 031, 32180”, “0002057074 031, 25458” and “0002070514 031, 38898”. These are passive tags containing 1 bit of information. Inclusion of lot of data is not possible on this small storage tags, for practical purpose we need more storage tags. Hence when we will take it near to the reader it will only identify the tag, and then check from the database as to which information is stored corresponding to this card number. The reader used is GP20-22. This reader works 12V which is provided from separate adapter, and is connected to the computer through a RS232 nine pin port, this port is discussed in detail further in this paper. The application software whose work is to connect to the reader through the serial port, gathers from the reader about the tag which is in contact at the instant. The software is made in VB .NET which also linked with the Microsoft Access Database, to get the information about the student, book and the date of issue and return. VB.NET is used because serial connectivity is easily achieved.

9.5. Explanation of its Working with Example

We take an example to explain the working in detail, suppose a person named "XYZ" comes to issue a book "ABC". He selects the book from the rack and brings it near to the reader, this on identification of the book will beep and on the screen a screen will pop up which list the information about the book from the database corresponding to the tag number associated with it. The read speed of the readers is very fast and has to be in contact with the reader for very short time. We will then have to add the information about the student XYZ in the text fields and click on the issue button. This manual information can be removed if we have separate ID card, containing information about student and can be kept with the book and this will be read simultaneously. For simultaneous reading we will have to add to our application software an *"Anti-Collision Theorem"* so that no two tags interfere with each other. Making the work simple we have only tagged the books to ease the understanding of the case study for the readers.On return of the book when the book is scanned near the reader then the information about the book, student and date of return is retrieved which also calculates the fine if any.

We have used three pins of the RS232 port, and interfaced it with software with RS232.dll file by placing it in location "*C:\WINDOWS\Microsoft.NET\Framework\ v1.1.4322\Rs232.dll*". We connect the database file named LibInfo.mdb by the command "*Dim conc As String = "Provider= Microsoft.Jet.OLEDB.4.0; Data Source = c:\LibInfo.mdb"*". The reader range is thatof about 10cm which is also the benefit of RFID system as it provides the flexibility of refraining from being in line of sight as compared to that of barcode.

10. CONCLUSION

Given the increasing use of RFID technology for a variety of purposes and applications, some of which with huge data protection implications [4].Various measures for high data security are been taken with the upcoming new crypto-graphical techniques. Efforts are there to decrease the power consumption by using GaAs and BiCMOS in tags to reduce the power consumption and the high cost. Various countries have also implemented RFID in their passports, credit cards and various other high data protection applications. We have given an overview of the various components used in the RFID system, discussed in detail about Tags and Reader and about the most important issue of Data Security. There is also the roadmap how this technology developed and how it works. The applications, advantages disadvantages have been also discussed here.

REFERENCES

[1] Klaus Finkenzeller, "*RFID Handbook*".

[2] Yang Guohao, Tian Jun and Chen Guochong, *"RF Controller Development and Its Application in Intelligent Transport System".*

[3] *"RFID Technology for Warehouse & Distribution Operations"*, LXE Inc., vol. no. 1-800-664-4593.

[4] *Official Journal L* 281, 23.11.1995, p. 31, available at: http://europa.eu.int/comm/internal_market/privacy/law_fr.htm

[5] Ann Cavoukian, *"Tag, You're it: Privacy Implications of Radio Frequency Identification* (RFID)"

[6] http://en.wikipedia.org/wiki/RS-232

[7] Atmel Corporation (1998) *Asset Identification EEPROM,* AT24RF08, San Jose, CA, http://www.atmel.com

[8] EC (1995) *the Radio Equipment and Telecommunications Terminal Equipment Directive* (1999/5/EC), http://europa.eu.int/comm /enterprise/rtte

[9] EAN.UCC (1999) EAN.UCC White Paper on Radio Frequency Identification, EAN International & UCC Inc., November, http:// www.ean-int.org

[10] EAN.UCC (2000) RFID and the EAN.UCC System, GTAG Project Team, EAN International & UCC Inc., http://www.eanint.org

[11] ERC (2000) CEPT marking and the R&TTE Directive, ERC Report 84, European Radio communications Committee (ERC), Lisbon, June, http://www.ero.dk

[12] Escort Memory Systems (1998a) *RFID Application — Case Study,* Automotive Engine Manufacturer, General Motors, Escort Memory Systems, Scotts Valley, California.

[13] Escort Memory Systems (1998b) *RFID Application — Case Study, Agricultural Equipment Manufacturer,* John Deere Company, Escort Memory Systems, Scotts Valley, California.

[14] Geers, R., Puers, B., Goedseels, V. and Wouters, P. (1997) *Electronic Identification, Monitoring and Tracking of Animals, CAB International,* Wallingford, ISBN 0-85199-123-8.

[15] Hanex (n.d.) *Sales presentation, Hanex RFID-System for Metal, HXID-System,* Hanex Co. Ltd., Japan.

[16] awkes, Peter (1997) *singing in Concert — some of the possible methods of orchestrating the operation of multiple RFIDTags enabling fast, efficient reading without singulation,* Amsterdam, 19 February.

[17] Kraus, John D. (1988) *Antennas*, 2nd edn, McGraw-Hill, New York, ISBN 0-07-100482-3.

[18] Kraus, Gunthard (DG8GB) (2000) *Moderner Entwurf von Patch-Antenna*, Part 1: UKW-Berichte,3; Part 2: UKW-Berichte 4, http://www.ukw-berichte.de

[19] Lee, Youbok (1999) *Antenna circuit design,* AN710, application note, microID 13.56MHz —RFID system design guide, Microchip, http://www.microchip.com

[20] Mansukhani, Arun (1996) *Wireless Digital Modulation, Applied Microwave & Wireless, November/December*

[21] Panasonic (n.d.) Technical Data Sheet — *Features of ferroelectric nonvolatile memory.*

[22] Philipp, Stefan (2001) CISC vs. *RISC and a plea for peace. Enhanced Microcontroller Architecture for Smart Card ICs,* Philips Semiconductors,http://www.semiconductors.philips.com/identification.

[23] Philips Components (1994) Datasheet, Ferrite roof antennas for RF-identification transponders, http://www.regtp.de

[24] Roz, Thierry and Fuentes, Vincent (n.d.) *Using low power transponders and tags for RFID applications,* Firmenschrift, EM Microelectronic Marin, CH-Marin.

[25] Schurmann, Josef (1993) *TIRIS — Leader in Radio Frequency Identification Technology*, Texas Instruments Technical Journal, November/December.

[26] Semiconductors Gratkorn GmbH and A-Gratkorn, http://www. semiconductors. philips. com/identification.

[27] Tag Master (1997) Datasheet: Mark Tag S 1255, *multiple access read-only-cards,* Tag Master AB, S-Kista.

[28] Texas Instruments Deutschland GmbH (1996) *Standard Transponder Specifications,* 06/1996.

[29] Vcd, Krebs D., Unpublished manuscripts, *Venture Development Corp.,* http://www.vdc-corp.com

In: Radio Frequency Identification ISBN 978-1-61122-416-0
Editor:Alison R. McAdams, pp. 125-139© 2011 Nova Science Publishers, Inc.

Chapter 7

COULD RFID-BASED SYSTEMS BE REGARDED AS MULTI-AGENT SYSTEMS?*

H. K. Chan†

Department of Industrial and Manufacturing Systems Engineering; Room 824, Haking Wong Building, University of Hong Kong, Pokfulam, Hong Kong

ABSTRACT

Radio Frequency Identification (RFID) technology has received overwhelming attention in recent years. RFID tags could be regarded as ubiquitous or pervasive computing devices, which can send or receive radio frequency signal and can store a number of data. It is in fact not a new technology because its first military application can be found in the Second World War. Recently, RFID-based systems are linked to multi-agents systems such that RFID-based devices are considered as agents, in particular mobile agents. However, reported literature is related to, relatively, high-level conceptual applications of RFID technology in agent-based systems. The answers to the following questions are still unclear: whether RFID-based systems could be represented as multi-

---

* A version of this chapter was also published in *Expert Systems Research Trends,* edited by A. R. Tyler, published by Nova Science Publishers, Inc. It was submitted for appropriate modifications in an effort to encourage wider dissemination of research.

† H. K. Chan: Phone: (852) 2859 7967 ; Fax: (852) 2858 6535 ; E-mail: hkchan@ieee.org

agent systems or not? If this is not true, what are the missing components in RFID-based systems in contrast to multi-agent systems? In other words, how RFID-based systems could be enhanced to become multi-agent systems? The objective of this chapter is to address these questions.

Keywords: RFID, agent, identification, pervasive

I. INTRODUCTION

Radio-frequency identification (RFID) technology is in fact not a new technology. It was used for aircraft identification during the World War II (Glidden et al., 2004). Since RFID-based systems are wireless in nature, non-contact reading (more specifically, no line of sight requirement) is possible and are effective in manufacturing systems which subject to some unfavorable environments. Therefore, RFID technology can provide additional features and removes boundaries that limited the use of previous alternatives (Want, 2004). However, industrial application of RFID technology is still in its infancy. As a matter of fact, the technology provides better and more accurate information, which makes it possible to know exactly where every item at any moment in a real time basis.

More specifically, the technology is "an automatic way to collect product, place, time, or transaction data quickly and easily without human intervention or error" (Bansai, 2003). RFID is similar in theory to barcode identification system. If we only compare RFID with barcode, which is one of the widely adopted identification technology nowadays, in terms of the ability of identification, the former definitely can outperform what the latter can do as reflected by the characteristics of RFID technology in later discussion. Everything barcode can do, RFID can do and the ability to do the tasks is even better.

Recently, this technology is linked with multi-agent systems in several applications like location-tracking systems (Satoh, 2004), intelligent guided-view systems (Chao, 2005), etc. Although RFID-based systems share some similarities to multi-agent systems, these studies concerned high-level application and did not discuss how RFID-based systems could be transformed into multi-agent systems. In this connection, this chapter sets out to explore how RFID-based systems could be mapped to multi-agent systems, and to find out what are the missing components in RFID-based systems with respect to multi-agent systems.

The rest of the chapter is organized as follows: Section II presents some basic ideas of the RFID technology, together with the benefits it can offer and its limitations. Section III discusses main characteristics of agents and multi-agent systems. Section IV then compares RFID-based systems and multi-agent systems. Finally, conclusion is presented in Section V.

II. Review of the RFID Technology

A. Basics of RFID

Technologically speaking, RFID technology is quite simple. Generally, RFID-based systems consist of three core components: (1) A small electronic data carrying device called a tag (or a transponders – transmitter and responder); (2) A reader or a scanner that communicates with the tag by using radio frequency signals; and (3) A host data processing system that contains information on the identified item and distributes information to other remote data processing systems (Keskilammi et al., 2003). RFID-based systems become complicated if more tags and readers are presented, and they are interconnected by different host systems, which may be located at different locations (and even different countries). In such a case, effective coordination among the devices so that resources are allocated properly is crucial to the successful implementation of the systems.

The simplest form of RFID-based system as an identification system consists of a tag and a reader, each equipped with an antenna (Bansai, 2003). The reader sends out Radio Frequency (RF) signal that forms a magnetic field. The RFID tag can "couple" the field by its antenna and the chip on the tag then demodulates the signal and responds to it accordingly. Afterwards, the tag may send back an RF signal to the reader, if necessary. Finally, the reader converts the signal into digital data. In most cases, the data is transferred to a host computer.

RFID-based systems work at a number of different frequencies: low frequency (125 kHz or 134 kHz), high frequency (13.56 MHz), and the emerging ultra-high frequency (300 MHz to 1 GHz) (Walko, 2004). The low frequency range tends to be used in chips for cars as well as asset management (Dorgenroth and Fobes, 2004). The high frequency range can be used for item-level management (for example, books, libraries and inventories). It may also be used in supply chain environments for goods on a conveyor belt and has a typical read/write distance of 1.5 meters (Dorgenroth and Fobes, 2004). The 13.56 MHz band is ideal for low-cost RFID-based systems because reader output power is maximized based on FCC specifications (Finkenzeller, 2003). The Ultra-High Frequency (UHF) range is being used in supply chain applications such as identification of pallets as well as individual boxes on pallets (Dorgenroth and Fobes, 2004). The main disadvantage of using this frequency range is that you have to occupy a frequency band, which is not the same in different countries. This non-standardized operating frequency is certainly a barrier to the interoperability of tags in different countries when a product moves from one country to another country (Glidden et al., 2004).

RFID tags can be broadly categorized into two groups: active and passive. The former is equipped with its own power source (e.g. battery) to transmit information directly to a reader (Caton, 2004). On the other hand, a passive tag can power up by electromagnetic induced energy from a source (e.g. a reader) through a tag's antenna, which is essentially a coil, and using that energy to transmit stored data to the reader, or to perform other operations (Caton, 2004). As microelectronic technology is getting more advanced and more sophisticated, the cost of a passive RFID chip is decreasing to as low as 50 cents (Walko, 2004). Therefore, passive RFID-based systems are easy to apply in manufacturing and logistics control systems because of its inexpensive cost (Keskilammi et al., 2003). In addition, the size of a passive RFID chip is just tiny, which is small and easy to attach to the object to be identified. These passive devices are suitable for use in a low cost RFID-based system operating in the 100 kHz to 15 MHz (Redinger et al., 2004).

B. Pros and Cons of RFID

Many reports in the literature have discussed the benefits and limitations of RFID technology. Table 1 and Table 2 summarize the benefits and limitations of RFID technology respectively. However, it should be noted that the scope of these two tables are limited to the technology of RFID technology itself, not the benefits or limitations of any particular RFID-based systems.

Table 1. Benefits of RFID technology

Benefits	Discussions
• No line of sight scanning	• Tags can be read without being visible to the reader, as long as they are exposed to the coverage of a reader.
• Flexibility	• Since there is no line of sight requirement as discussed above, tagged products can be packaged in, for example, harsh fluid and chemical environments and can be read in rough handling situations.
• Long Reading Range	• Tags can be read over very long range, especially those systems with UHF frequency tags. This attribute is superior in mass logistic applications which need a range of at least a metre and up to 4 or 5 metres.
• High speed of reading a tag	• The speed of reading a tag is very fast. This is especially useful when the items needing to be identified are moving quickly, for example, on a conveyor in a production system.
• Ability to read multiple items	• A number of tagged items can be read at the same time in a short period of time.
• Durability	• Performance of paper-based barcodes can be degraded easily subject to the environment, e.g. if they become wet. This is not an issue that affects RFID tags.
• Programmability	• Many tags are read / write capable, rather than read only, like barcodes. In addition, tags with different memory sizes are available for various applications which may need extra functionality. Of course, additional cost is incurred for tags with higher memory size.

Source: Chan and Chan, 2007.

From Table 1, it is obvious that RFID can not only replace traditional barcode technology, it also provides additional features and removes boundaries that limited the use of previous alternatives (Want, 2004). On the other hand, among the limitations of RFID technology as listed in Table 2, cost and standardization are the most important issues to be addressed in applications of RFID in the industry.

By looking at the characteristics, benefits, and limitations of RFID-based systems, a short conclusion can be drawn for RFID-based systems: they are in fact another type of information system that makes use of radio wave to collect data. Regardless of the benefits it can offer and limitations we need to consider, RFID is only useful if a proper tool is used to analyze the large amount of collected data. Software solutions have to be developed to cope with the huge volume of data that RFID-based systems can generate because the infrastructure from suppliers is often incompatible (Walko, 2004). In addition, if the technology is only utilized for in-house applications, the cost impact will certainly be minimized if tags are designed as reusable. Also, non-standardization of RFID operations (e.g. frequency) would not be a hurdle anymore since the applications are not required to be used somewhere outside the factory.

Table 2. Limitations of RFID technology

Limitations	Discussions
• Cost	• Since the unit cost of a RFID tag is still high, it is still not appropriate to use RFID tags as a direct replacement of barcodes without any other value added business benefits.
• Reliability	• RF signal may be absorbed by moisture, be it in the product or the environment, and may be distorted or even absorbed by metal. This means that tags might be unable to be read in some circumstances.
• Tag and readers collisions	• Tag collisions occur when two tags try to respond to the reader at the same time while reader collisions occur when two neighboring readers interrogate a tag simultaneously and confuse it, or when they "blind" each other to the relatively weak tag responses (Sarma et al., 2001). These problems require anti-collision mechanisms to coordinate behavior between tags and readers.
• Electrical interference	• RF transmission and communications may interference by electronic noise (e.g. fluorescent light or electric motors).
• Accuracy	• RFID is not yet a 100% solution (Anderson, 2004). It can be difficult to identify and read a specific tag from all the others that are within the range of a reader.
• Standard	• There is a lack of standard govern RFID operations. An open system is sough in order to ensure interoperability while the tags are moving from one place to another place, or even one country to another country with the products.
• Security	• The ability to write information to tags is one of the main benefits of RFID technology. The mechanism, however, needs to be secure to ensure that unauthorized parties are unable to write false information into the tag. In addition, competitors may be able to retrieve sensitive information from the tagged products. Therefore, security is an issue to be tackled before an application can be phased in.
• Privacy	• Some consumer groups have concerns about the possibly privacy implications of RFID (Weis 2004).

Source: Chan and Chan, 2007.

III. CHARACTERISTICS OF MULTI-AGENT SYSTEMS

Research in "Multi-Agent Systems" is a prevalent topic in the last two decades in many areas. It is a branch of Artificial Intelligence called Distributed Artificial Intelligence, which attempts to compensate for the deficiencies of classical Artificial Intelligence with regard to the development of intelligent agent. An agent can be defined as "a computer system, situated in some environment that is capable of flexible autonomous action in order to meet its design objectives" (Jennings et al., 1998). In a multi-agent system, a number of heterogeneous agents are working independently, or in a cooperative and interactive manner to solve problems in a decentralised environment.

There are number of factors that could make agent-based approach applicable for problem solving (Wooldridge, 2002). They are summarized as follows:

i. To be applied in open environments, or at least highly dynamic, uncertain, or complex environments, because systems which are capable of flexible autonomous action are often the only solution.
ii. To be applied in systems which require the units or objects in which either cooperating with each other to solve complex problems, or else competing with one another. Many of such scenarios can be modeled as multi-agent systems.
iii. To be applied in decentralized environment. In such situations, distribution of either data, control, or expertise means that a centralized solution is difficult, if not impossible. Such systems may often be modeled as multi-agent systems easier than the centralized counterpart.
iv. Agents can act as a mediator or wrapper so that some legacy systems or software can be retained. As a mediator, agents can interact with other software components. As a wrapper, the legacy systems are "wrapped" by agents, which enabling them to interact with other software components.

The main characteristics of an agent, by definition, are that it is reactive, goal-oriented, able to learn, autonomous, mobile, and able to communicate with other agents (Brenner et al., 1998). It was suggested in Wooldridge and Jennings (1995) that an agent should possess the following capabilities:

i. Autonomy – agents operate without the direct intervention of humans or others, and have some kind of control over their actions and internal state;
ii. Reactivity – Agents are able to sense any changes in their environment, and respond in a timely fashion to changes that occur in it in order to satisfy their design objectives;
iii. Proactiveness – Agents are goal-directed such that their behaviors are taking the system objectives into consideration;
iv. Social ability – Agents can communicating with other agents to exchange information or to make decisions collectively in order to satisfy their design objectives.

Mobile agents are specialized agents that are not bound to the system where they begin execution but are free to travel through a network (Ghiassi and Spera, 2003). This distinctive feature enables mobile agents to transport freely from one system to another. Specifically, mobile agents are "independent programs or mobile codes that can travel from one network to another while performing different kinds of operations", and "can be initiated on a single host and then migrate from host to host over a network" (Gupta et al., 2001). Mobile agents can move locally to the resources that they need, requiring neither the transfer of data nor network connection during the access (Omicini and Zambonelli, 1999). A key issue that hinders mobile agent applications and needs to be addressed for mobile agent systems is administrability of mobile agent systems through, for example, authorization policies (Luck et al., 2004). If a mobile agent wants to "talk" to other agents, or to acquire information or knowledge that it needs from a local source without network communication, it can move to a system that contains the agents freely (Shin and Jung, 2004).

Although there is plenty of literature that could be found in regard to architectural issues of multi-agent systems, or collaborative control in agent-based systems, only a few of them have considered the implementation of agent-based systems in a real-time basis. RFID technology is a potential candidate for deploying such systems. In the next section, comparison of RFID-based systems with multi-agent systems will be made according to these definitions of agents and agent-based systems, and the appropriateness to apply multi-agent systems for modeling RFID-based systems.

IV. Relationship between RFID-Based Systems and Multi-Agent Systems

A. Recent Development of Agent-Oriented RFID-Based Systems

Recently, a number of agent-oriented RFID-based systems have been proposed. Most of them were based on the mobile nature of RFID tags in order to enable some innovative ideas or concepts. For example, Satoh (2004) presented a general-purpose framework for RFID-based location tracking systems. The framework can navigate Java-based mobile agents to stationary or mobile computers near the location so of the entities and places to which the agents are attached, even when the locations change. RFID tags are attached to a person, place, or thing in the physical world and they are used to detect the locations of such physical entities in a building and enable applications to respond to these locations. In other words, Satoh treated an object with RFID tags as mobile agents which movement are monitored by a "Mobile Agent Navigation" module in the framework. Chao (2005) designed an intelligent guided-view system based on RFID technology. The system makes use of RFID technology to serve as the nerve ending, which explores the environment information around user and collect the user pattern, for the backend artificial intelligent systems like expert system to analyze all the input information. In the system, RFID tags are also regarded as mobile agents. These extended applications in area locating systems highlight the advantages of RFID technology (e.g. scanning without line of sight).

In the scope of supply chain management, RFID technology is certainly an enabler for forming agent-based dynamic information network, such as gathering point-of-sales data (Ahn and Lee, 2004), control for mass customization manufacturing (Liu et al., 2004), inter-organization event management system to optimize supply chain performance (Zimmermann and Paschke, 2003), etc. It can foresee that more relevant literature will appear in the near future. Above literature, however, assumes RFID-based systems are multi-agent systems intrinsically. Although the two systems share some similarities, the former is not a direct subset of the latter. There is a need to explore whether RFID-based systems could be modeled by multi-agent systems or not. If so, one might ask how to transform RFID-based systems to multi-agent systems. These issues will be discussed in the following section.

B. Comparison between RFID-Based Systems and Multi-Agent Systems

Although a number of agent-based RFID applications have been reported, one might argue whether it is appropriate to model RFID-based systems as multi-agent systems or not. To answer this question, we can refer to the appropriateness to employ multi-agent systems as advocated by Wooldridge (2002) in Section III. The four characteristics are analyzed with respect to RFID-based systems as follows:

i. Open Environment – RFID-based systems could be operated in either open or closed systems with respect to a particular company. For example, if it is used to monitor the location of a batch of inventory during the course of transportation from one country to anther country, the environment is certainly an open system. However, even in closed systems, RFID tags are free to move along with the carriers (people or goods). In this regard, RFID-based systems are still operating in "open" systems;
ii. Agents are a natural metaphor – It is not obvious to see RFID-based systems fulfill this "prerequisite". This is because most RFID tags are memory devices and have no computing power. This leads to the fact that a society of "RFID tags" are not able to communicate and cooperate with each other to solve complex problems. In order to fill this gap, RFID-based systems must encompass with middleware in order to handle the information obtains from RFID tags. This middleware acts as a middleman to assist communications among RFID tags. We can call this middleware as mediated agent as quoted in many reported literature;
iii. Distribution of data, control or expertise – Since each RFID tag carries its own information, and is a distributed entity throughout its application life cycle, a centralized solution seems impossible to monitor and collect information among RFID tags. Therefore, RFID-based systems involves distribution of data, control and expertise;
iv. Legacy systems – Theoretically, an RFID-based system could be a brand new system such that any legacy systems could be discarded. However, it is not economical to do so because RFID-based systems are just one component of the whole enterprise systems. The wrapper solution that agent-based systems could offer enable information

obtains from RFID tags to be incorporated with other existing software components.

Through this analysis, it is uncontroversial to say that it is pretty "appropriate" to model RFID-based systems as multi-agent systems, as long as a proper middleware exists. Then, one might ask how to model RFID-based systems as multi-agent systems. The fundamental characteristics of RFID-based systems are certainly not specific enough to be recognized as multi-agent systems. By looking at the characteristics of multi-agent systems as suggested in Wooldridge and Jennings (1995) in Section III, some insights could be gained on how to turn RFID-based systems to multi-agent systems and they are explained as follows:

i. Autonomy – To a certain extent, this is one of the characteristics of RFID technology. This is because RFID tags could be read remotely without human intervention. However, most of the RFID tags are just memory devices without computing power. They are not able to control over their actions (e.g. write or change its information), in spite of the fact that they are not designed to do so. Nevertheless, it is easy to "enable" RFID tags to fully possess this attribute by assigning unique identification number to each tag under the system so that mediated agents could be able to take action on behalf of the tags (e.g. instruct a tag to write or to change information according to the environment that it is exposed to in different stage of manufacturing processes). IPv6 may be a possible solution to assign unique identification number to each tag because there are more than enough IP addresses in the new version and they are reusable according to the design of the systems anyway. In such case, communication among RFID devices could follow the TCP/IP protocol;
ii. Reactivity – RFID tags are only able to respond to readers (another agent in the RFID-based systems), which are located in proximity to them. In other words, the tags themselves are not able to perceive their environment, and respond in a timely fashion to changes that occur in it. In other words, an RFID-based system possesses this attribute only if a network of readers is working in a cooperative manner by "telling" the RFID tags what are the environment they are exposed to. If so, the agents in the systems could "perceive" their environment with the help of other agents (i.e. RFID readers);

iii. Proactiveness – Since the major design objective of RFID-based systems is for identification. Therefore, this characteristic is intrinsically fulfilled by the design of RFID tags since they can be read and written by RFID readers according to the goal of the system they belongs to. Of course, mediated agents (i.e. middleware) are more dominant in taking the initiative in order to satisfy the ultimate goal of the system (e.g. accurate inventory replenishment);

iv. Social ability – RFID tags are designed to communicate with RFID readers. Therefore, with a proper readers network as suggested above, RFID tags are "virtually" capable of interacting with other agents (tags and humans).

In short, active RFID devices play a vital role in enhancing a RFID-based system to form a multi-agent system. The following table summarizes the above discussions:

Table 3. Relationships between RFID-based systems and multi-agent systems

Attributes of Multi-agent Systems	How RFID-based Systems can have Possession of these Attributes
• Autonomy	• Active RFID devices have to be used in a network as mediators in order to couple with passive RFID tags, which are assigned a unique ID in the network. Otherwise, RFID tags are only memory devices which could not exhibit autonomy.
• Reactivity	• The RFID-based systems should be designed in the way that a reader, or an active RFID device, should be coupled with a number of passive RFID devices so that the former can "inform" the environment of the passive devices and appropriate actions can be taken. In other words, the active devices react to the environment and the passive devices react to the active devices.
• Proactiveness	• This is already an intrinsic characteristic of any RFID-based systems. However, mediators in the systems could further improve the proactiveness of the systems.
• Social ability	• Communication among RFID devices is a major benefit of any RFID-based systems and thus this attribute, like proactiveness, is fulfilled inherently.

As reflected from the discussions above, RFID-based systems could be modeled as multi-agent systems easily, with the aids of a proper design of RFID readers network and middleware. However, if the systems span across different companies, a consensus of the read and write capability should be defined clearly. Otherwise, the aggregate readers network becomes individual sub-networks, which own by different companies. This would certainly reduce the synergistic benefits that the whole systems could deliver.

V. CONCLUSION

It cannot be denied that every tagged item has the potential to become an intelligent and communicative agent. However, it is actually not the tag or reader that is the significant development, but rather the information itself — information about the location and status of goods worldwide, which will be available to manufacturers, distributors, and retailers simultaneously — that makes RFID an enabling technology. Of course, it is not necessarily to represent or model RFID-based systems as multi-agent systems. However, agent technology is a fast development area in the E-commerce domain so that if RFID-based systems could be developed as multi-agent systems, synergistic benefits could be gained. In such applications, RFID tags could be regarded as smart devices, which are equipped with computing resources (together with readers and middleware, of course) and "able to communicate with other similar objects via any physical transmission medium" (Carabelea and Boissier, 2003). It is interesting in uplifting RFID technology with multi-agent systems in the future.

REFERENCES

Anderson, C. (2004). EVERYTHING you always wanted to know about RFID... but were afraid to ask. *Logistics Management*, vol. 43, no. 9, pp. 57-60.

Ahn, H. J., and Lee, H., (2004). An agent-based dynamic information network for supply chain management. *BT Technology Journal*, vol. 22, no. 2, pp. 18-27.

Bansai, R. (2003). Coming Soon to a Wal-Mart Near You. *IEEE Antennas Propagation Magazine*, vol. 45, no. 6, pp. 105-106.

Brenner, W., Zarnekow, R., and Wittig, H. (1998). *Intelligent Software Agents: Foundations and Applications*. Berlin: Springer.

Carabelea, C., and Boissier, O. (2003). Multi-agent Platforms for Smart Devices: Dream or Reality? In *Proceedings of the Smart Objects Conference (SOC'03)*, Grenoble, France, May 2003, pp.126-129.

Caton, M. (2004). RFID reshapes the supply chain. *eWeek*, vol. 21, no. 16, pp. 45-49, April 19.

Chan, H. K., and Chan, F. T. S. (2007). Is the RFID Technology Ready To Integrate Supply Chain Activities? *International Journal of Enterprise Network Management*, in press.

Chao, H. P. (2005). The Non-specific Intelligent Guided-View System Based On RFID Technology. In *Proceedings of the International Conference on Advanced Information Networking and Applications (AINA'05)*, 25-30 March, 2005, vol. 2, pp. 580-585.

Finkenzeller, K. (2003). *RFID Handbook: Fundamentals and Applications in Contactless Smart Cards and Identification* (2nd ed.). New York: Wiley.

Ghiassi M., and Spera, C. (2003). Defining the Internet-based supply chain system for mass customized markets. *Computers and Industrial Engineering*, vol. 45, no. 1, pp. 17-41.

Glidden, R., Bockorick, C., Cooper, S., Diorio, C., Dressler, D., Gutnik, V., Hagen, C., Hara, D., Hass, T., Humes, T., Hyde, J., Oliver, R., Onen, O., Pesavento, A., Sundstrom, K., and Thomas, M. (2004). Design of Ultra-Low-Cost UHF RFID tags for Supply Chain Applications. *IEEE Communication Magazine*, vol. 42, no. 8, pp. 140-151.

Gupta, A., Whitman L., and Agarwal, R. K. (2001). Supply Chain Agent Decision Aid System (SCADAS). In *Proceedings of the 2001 Winter Simulation Conference*, pp. 553-559.

Jennings, N. R., Sycara K., and Wooldridge, M. (1998). A Roadmap of Agent Research and Development. *Autonomous Agents and Multi-Agent Systems*, vol. 1, no. 1, pp. 7–38.

Keskilammi, M., Sydänheimo L., and Kivikoski, M. (2003). Radio Frequency Technology for Automated Manufacturing and Logistics Control. Part1: Passive RFID Systems and the Effects of Antenna Parameters on Operational Distance. *International Journal of Advanced Manufacturing Technology*, vol. 21, no. 10-11, pp. 769-774.

Liu, M. R., Zhang, Q. L., Ni, L. M., and Tseng, M. M. (2004). An RFID-Based Distributed Control Systems for Mass Customization Manufacturing. In *Proceedings of the International Symposium on*

Parallel and Distributed Processing and Applications (ISPA), pp. 1039-1049.

Luck, M., McBurney, P., and Preist, C. (2004). A Manifesto for Agent Technology: Towards Next Generation Computing. *Autonomous Agents and Multi-Agent Systems*, vol. 9, no. 3, pp.203-252.

Morgenroth D., and Fobes, K. (2004). Another link in the chain. *Card Technology Today*, vol. 16, no. 4, pp. 11-12.

Omicini, A., and Zambonelli, R. (1999). Coordination for Internet Application Development. *Autonomous Agents and Multi-Agent Systems*, vol. 2, no. 3, pp.251-269.

Redinger, D., Molesa, S., Yin, S., Farschi R., and Subramanianm, V. (2004). An Ink-Jet-Deposited Passive Component Process for RFID. *IEEE Transactions on Electron Devices*, vol. 51 no. 12, pp. 1978-1983.

Sarma, S., Brock D., and Engels, D. (2001). Radio Frequency Identification and the Electronic Product Code. *IEEE Micro*, vol. 21, no. 6, pp. 50-54.

Satoh, I. (2004). A mobile Agent-based Framework for Location-based Services. In *Proceedings of the IEEE International Conference on Communication*, 20-24 June, 2004, vol. 3, pp. 1355-1359.

Shin, M., and Jung, M. (2004). MANPro: mobile agent-based negotiation process for distributed intelligent manufacturing. *International Journal of Production Research*, vol. 42, no. 2, pp. 303-320.

Walko, J. (2004). Super-locator. *IEE Communications Engineer*, vol. 2, no. 1, pp.10-13.

Want, R. (2004). Enabling Ubiquitous Sensing with RFID. *IEEE Computer*, vol. 37, no. 4, pp. 84-86.

Weis, S. A. (2004). RFID Privacy Workshop – Concerns, Consensus, and Questions. *IEEE Security and Privacy Magazine*, vol. 2, no. 2, pp. 48-50.

Wooldridge, M. (2002). *An Introduction to MultiAgent Systems*. Chichester, England: John Wiley and Sons.

Wooldridge M., and Jennings, N. R. (1995). Intelligent Agents: Theory and Practice. *Knowledge Engineering Review*, vol. 10, no. 2, pp. 115-152.

Zimmermann, R., and Paschke, A. (2003). PAMAS – An Agent-Based Supply Chain Event Management System. In *Proceedings of the Americas Conference on Information Systems*, pp. 1892-1900.

INDEX

C

D

E

F

G

H

I

S

T

U

V

W

X

Y